Microgrid Energy Management

Microgrid
Energy Management

Editor

Pietro Varilone

MDPI • Basel • Beijing • Wuhan • Barcelona • Belgrade • Manchester • Tokyo • Cluj • Tianjin

Editor
Pietro Varilone
University of Cassino and Southern Lazio
Italy

Editorial Office
MDPI
St. Alban-Anlage 66
4052 Basel, Switzerland

This is a reprint of articles from the Special Issue published online in the open access journal *Energies* (ISSN 1996-1073) (available at: https://www.mdpi.com/journal/energies/special_issues/MEM).

For citation purposes, cite each article independently as indicated on the article page online and as indicated below:

LastName, A.A.; LastName, B.B.; LastName, C.C. Article Title. *Journal Name* **Year**, *Volume Number*, Page Range.

ISBN 978-3-0365-1711-7 (Hbk)
ISBN 978-3-0365-1712-4 (PDF)

Contents

About the Editor

Pietro Varilone received his M.Sc. degree in Electrical Engineering from the Università di Cassino, Italy, in 1995, and his Ph.D. degree in Electrical Energy Conversion at the Second University of Napoli (Italy), Italy, in 1999. He is currently Associate Professor of Power Systems at the "Maurizio Scarano" Department of Electrical and Information Engineering of the University of Cassino and Southern Lazio (Italy). He is a member of the Ph.D School Board of "Methods, Models and Technologies for Engineering" and the scientific director of the "LaSE" Power System Laboratory of the University of Cassino and Southern Lazio, (Italy). His current research includes the analysis of transmission and distribution in unbalanced systems, power quality issues—with a particular focus on voltage sags—and the development of distribution networks towards Smart Grids. He has authored more than 100 scientific papers and some chapters in international books on power system planning and operation.

Preface to "Energies Microgrid Energy Management"

In IEEE Standards, a Microgrid is defined as a group of interconnected loads and distributed energy resources with clearly defined electrical boundaries, which acts as a single controllable entity with respect to the grid and can connect and disconnect from the grid to enable it to operate in both grid-connected or island modes. Together with distributed resources, such as renewable energy power plants and storage systems, and loads that actively contribute to the operation of such systems, an increasingly complex and performing ICT-relevant infrastructure characterizes these new electrical systems. A Microgrid must meet several needs and expectations from customers and from the various stakeholders involved in the electrical energy chain, with the consequence that optimal energy management is required to address a variety of different targets related to efficiency, power quality, resiliency, and affordability; moreover, new technologies and services have to be introduced to obtain these targets. Energy management systems should also carefully operate distributed energy resources and loads to avoid line overloading and critical voltage profiles inside the Microgrids. In this context, this Special Issue focuses on innovative strategies for the management of the Microgrids and, in response to the call for papers, six high-quality papers were accepted for publication. Consistent with the instructions in the call for papers and with the feedback received from the reviewers, four papers dealt with different types of supervisory energy management systems of Microgrids (i.e., adaptive neuro-fuzzy wavelet-based controls, cost-efficient power-sharing techniques, and two-level hierarchical energy management systems); the proposed energy management systems are of quite general purpose and aim to reduce energy usages and monetary costs. In the last two papers, the authors concentrate their research efforts on the management of specific cases, i.e., Microgrids with electric vehicle charging stations and for all-electric ships. I would like to thank the Energies journal for hosting this important topic and the press production team for their support along the way. We also thank the authors for their valuable contributions and the reviewers for their critical analysis and feedback. We do hope that the papers in the Special Issue will be a stimulus to all people working in the Microgrids area to understand the need for integrating theoretical studies with field applications and to understand the difficulties associated with the practical implementation of energy management systems.

Pietro Varilone
Editor

Article

Novel Improved Adaptive Neuro-Fuzzy Control of Inverter and Supervisory Energy Management System of a Microgrid

Tariq Kamal [1,2,*], Murat Karabacak [3], Vedran S. Perić [4], Syed Zulqadar Hassan [5] and Luis M. Fernández-Ramírez [2,*]

1. Department of Electrical and Electronics Engineering, Sakarya University, Faculty of Engineering, 54050 Serdivan/Sakarya, Turkey
2. Research Group in Sustainable and Renewable Electrical Technologies (PAIDI-TEP-023), University of Cadiz, Higher Polytechnic School of Algeciras, 11202 Algeciras (Cadiz), Spain
3. Department of Electrical and Electronics Engineering, Sakarya University of Applied Sciences, 54050 Serdivan/Sakarya, Turkey; muratkarabacak@sakarya.edu.tr
4. Munich School of Engineering, Technical University of Munich, 85748 Garching, Germany; vedran.peric@tum.de
5. Department of Electrical Engineering, Faculty of Engineering & Architecture, University of Sialkot, 51040 Sialkot, Pakistan; syedzulqadar.hassan.pk@ieee.org

* Correspondence: tariq.kamal.pk@ieee.org (T.K.); luis.fernandez@uca.es (L.M.F.-R.); Tel.: +90-536-6375731 (T.K.)

Received: 2 August 2020; Accepted: 8 September 2020; Published: 10 September 2020

Abstract: In this paper, energy management and control of a microgrid is developed through supervisor and adaptive neuro-fuzzy wavelet-based control controllers considering real weather patterns and load variations. The supervisory control is applied to the entire microgrid using lower–top level arrangements. The top-level generates the control signals considering the weather data patterns and load conditions, while the lower level controls the energy sources and power converters. The adaptive neuro-fuzzy wavelet-based controller is applied to the inverter. The new proposed wavelet-based controller improves the operation of the proposed microgrid as a result of the excellent localized characteristics of the wavelets. Simulations and comparison with other existing intelligent controllers, such as neuro-fuzzy controllers and fuzzy logic controllers, and classical PID controllers are used to present the improvements of the microgrid in terms of the power transfer, inverter output efficiency, load voltage frequency, and dynamic response.

Keywords: inverter; supervisory control; adaptive control; photovoltaic; ultra-capacitor; battery; wavelets; energy management

1. Introduction

Distributed generation (DG) systems based on renewable energy sources (RES), such as solar, wind, biomass, and hydropower, which are increasing steadily across the globe, are important in the generation of clean energy. In DG, energy conversion systems are placed near to the end consumers and large units are replaced with smaller ones. DG enables lower active power losses and operational costs, increased operational performance, and increased energy efficiency of the power system. Power system regulators are turning towards RES-based DG systems, along with the conventional centralized generation systems [1].

In DG and microgrid systems, one of the most critical parts is the inverter, because of its extensive range of functions [2]. Their operation is standardized by many international industrial standards and requirements such as IEEE 929-2000, EN61000-3-2, U.S. National Electrical Code 690, and IEEE 1547.

These standards describe some important parameters and properties of grid-coupled inverter, such as total harmonic distortion (THD), electromagnetic interference, voltage fluctuation, power quality, and power factor [3–5].

Owing to the expansion of DG and/or microgrid systems, many inverter designs and their control strategies have been published in the literature. For instance, fixed gain controllers (PI/PID) have been adopted by many researchers. For example, a PI controller with grid voltage feed-forward was used by the authors of [6,7], but some well-known drawbacks, such as a poor performance due to the integral action and the inability to track a sinusoidal signal, appeared in this method. These drawbacks have been addressed in the literature [8] by using a second order integrator. This approach is the most promising in terms of frequency synchronization, but the estimated frequency holds low frequency oscillations in case of DC offset being present in the grid voltage [9]. Similarly, in the literature [10], an adaptive control method was suggested using a direct current control scheme. The main drawback of the direct current control scheme is that there is no fixed systematic methodology to tune the PI controller, and therefore, an optimal direct current control is challenging to achieve.

Some researchers have preferred the applications of multilevel inverters (e.g., flying-capacitor and cascade H-Bridge neutral-point-clamped) in RES technologies [11–14]. However, the main problem of multilevel topologies is the unbalanced voltage between the capacitors across the DC link [15]. Similarly, numerous control strategies and algorithms on grid interactive inverters have been investigated and developed by different authors in past literature.

Some other techniques/controllers applied to inverters are predictive control [16,17], fuzzy control [18,19], sliding mode control [20], neural network (NN)-based control [21], and neuro-fuzzy (NF) [22]. All of the aforementioned techniques have their own advantages and drawbacks. For example, predictive control needs high computation efforts [23], while chattering limits the applications of the sliding mode control [24]. Fuzzy control has suffered of criticism for lacking a systematic strategy and a stability analysis technique. Similarly, in NN, each unit of the plant must be turned to produce control rules, and therefore its limitation is versatility [25]. Moreover, the NF method stays in the initial local minima during the search space [26]. This drawback in the existing NF controller motivates us to present a new controller based on the Jacobi wavelet. In the literature, many studies have shown that the use of a wavelet improves the performance of the NF network in RES [27–36].

Furthermore, an autonomous operation via distributed power sources improves its performance in terms of power sharing and voltage regulation. This operation can be obtained using energy management to supervise and control the power flow in the microgrid. For example, many authors have used different energy management systems for microgrids [37–39]. The authors of [37] controlled the microgrid via centralized management control, and therefore, the overall system could get away from communication when failure happened at a single location. The drawback of [38] was the absence of a countermeasure in the lower–top layers in its energy management strategy. Similarly, the researchers in [39] developed energy management of a microgrid using multiple-time optimization problems, but it could only give a day-ahead forecast, and power fluctuations and other related regulation schemes in the microgrid were not considered.

The main contributions of this research work are as follows:

(1) A supervisory energy management based on a two-level setting. The top-level controller determines set points for individual microgrid subsystems/components, i.e., a photovoltaic (PV), ultra-capacitor (UC), battery, and inverter, using weather statistics and load conditions. The lower level ensures that the top-level set points are accurately followed by the microgrid components. The operation of the microgrid is checked using real-world records of weather patterns and load fluctuations.

(2) A new adaptive controller based on the Jacobi wavelet neuro-fuzzy structure is developed for the grid coupled inverter, which yields a better THD, active and reactive power tracking, frequency, and efficiency than those achieved by other controllers that have appeared in the literature.

The rest of this research work is arranged as follows. The structure and control of the microgrid is given in Section 2. In Section 3, the supervisory algorithm is discussed. Section 4 provides the results and a comparison with the other existing intelligent and classical controllers. Finally, Section 5 draws the conclusions of this research.

2. Structure and Control of Microgrid

2.1. Structure of the System

Figure 1 illustrates the general structure of the proposed microgrid under analysis. It consists of a PV array, UC, and battery storage system. The PV array is capable of generating 261 kW under variable weather patterns and the UC/battery storage system are integrated for backup during excess power demand and as an energy storage system for surplus power. A DC–DC boost converter connects the PV array to the inverter. Similarly, two non-isolated buck boost converters connect the UC/battery to the DC bus, followed by the inverter. The inverter is controlled using a new wavelet-based adaptive controller. The adaptive-based controlled inverter is then coupled to the AC link at the grid and load. Energy management, power sharing, and transferring among PV, UC, and the battery with the rest of the microgrid are performed via energy management and supervisory control system (EMSCS).

2.2. Control of PV

The power of PV varies according to the weather patterns; therefore, its output is controlled by using an incremental conductance (IC) maximum power point tracking (MPPT) method, and is then regulated via a DC–DC boost converter, as illustrated in Figure 2. In PV, the optimal terminal voltage is determined by minimizing the MPPT error, which is shown as "e" in Figure 2, which is determined on a P/V curve through the IC method. The boost converter is controlled using PID controllers through the control of the duty cycle.

2.3. Control of Ultra-Capacitor/Battery

Both the UC and battery are controlled through PIDs embedded in DC–DC buck-boost converters. The boost mode permits power flow from the UC and/or battery to the DC during power demand bus, and the buck mode is utilized to charge the UC and/or battery from the DC bus during surplus power. Another advantage of the UC/battery is to regulate the DC bus voltage during abrupt weather variations and load changes. The control diagrams of the UC and battery are illustrated in Figures 3 and 4, respectively.

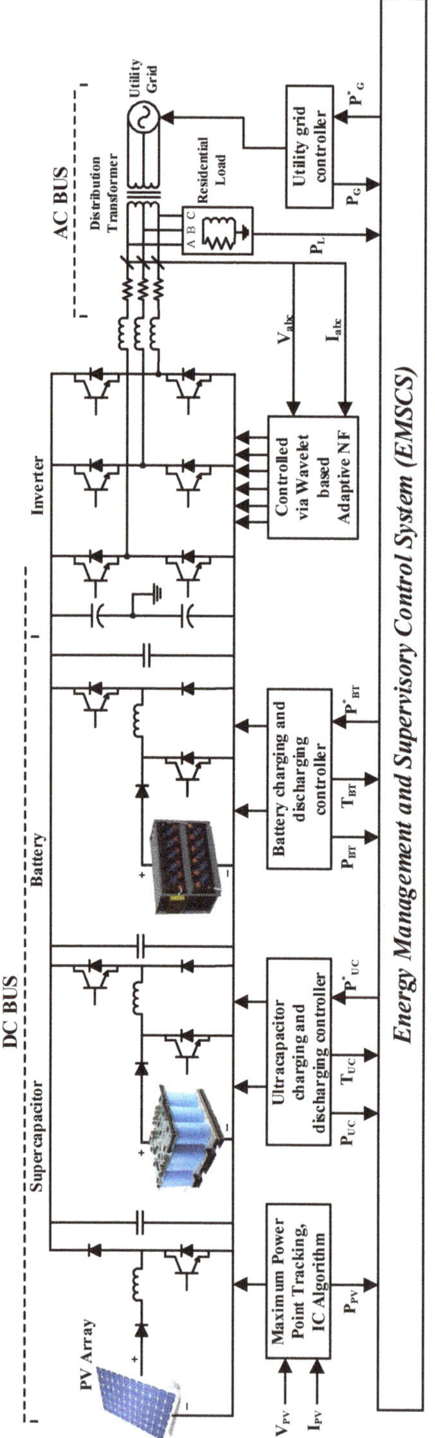

Figure 1. Structure of the proposed microgrid.

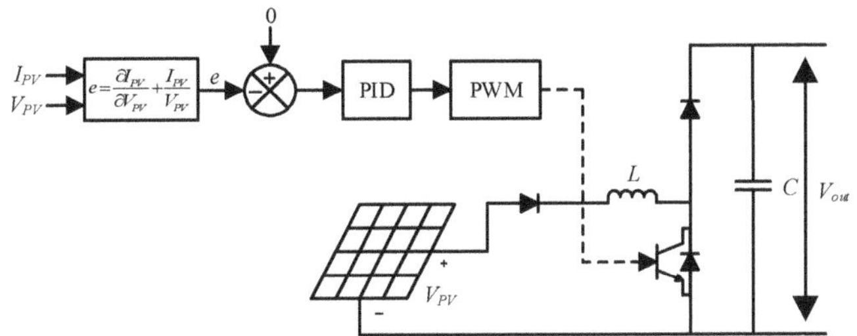

Figure 2. Control of photovoltaic (PV).

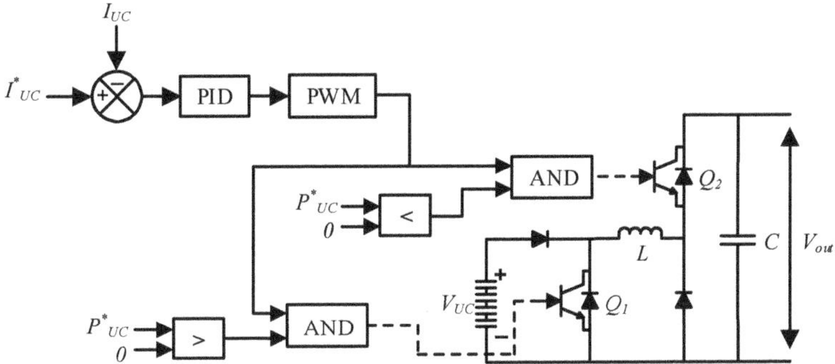

Figure 3. Control of UC.

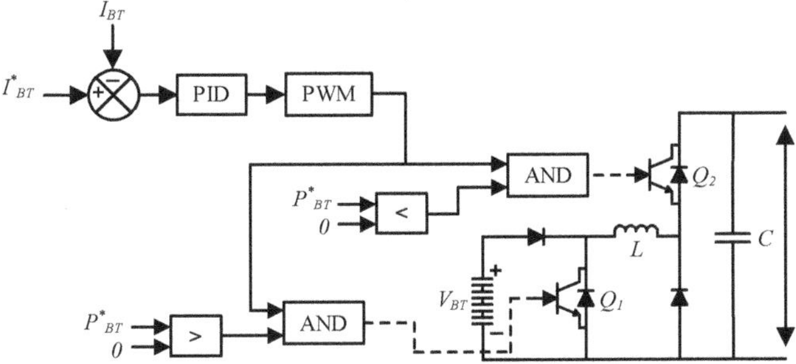

Figure 4. Control of the battery.

2.4. Control of Inverter

Inverters are critical for injecting power from the DG or microgrid into the grid or grid-coupled load. The control law for the proposed problem is written as follows:

$$
\begin{aligned}
& U_{inv}\Big|_{t\to\infty} = \left[y_{(P,Q)_{inv}}(t) \to y_{(P,Q)_{inv\_ref}}(t) \right] \\
& \max\ \eta_{(P,Q)_{inv}} = \frac{\int_0^t (P,Q)_{inv}(t)dt}{\int_0^t (P,Q)_{ref}(t)dt} \\
& \text{subjected to}: \\
& \qquad \Delta THD_{load} < \pm 5\% \\
& \qquad \Delta f_{load} < \pm 8\%
\end{aligned}
\tag{1}
$$

where P and Q stand for the active and reactive powers, respectively. According to the control law, the amount of P and Q injected into the load and grid from the inverter must be done according to the desired powers defined to the controller, provided that the maximum efficiency (η) is obtained while keeping the power quality constraints, i.e., THD and frequency (f), according to the IEEE standards [3–5]. The value of the power quality constraints must be kept smaller in order to reduce the power losses in the system.

In this work, the inverter was controlled through two adaptive neuro-fuzzy Jacobi wavelet (ANFJW)-based controllers—one controller used for controlling the active power, and another for controlling the reactive power delivered by the inverter to the grid. During operation, both the P and Q generated by the plant (microgrid) were compared with the references for P and Q. These differences were provided to the respective ANFJW controllers. Both ANFJW controllers operated on the difference and generated the corresponding reference currents (id* for active power and iq* for reactive power). Finally, the inverter switching commands/signals were generated using hysteresis current control method as shown in Figure 5.

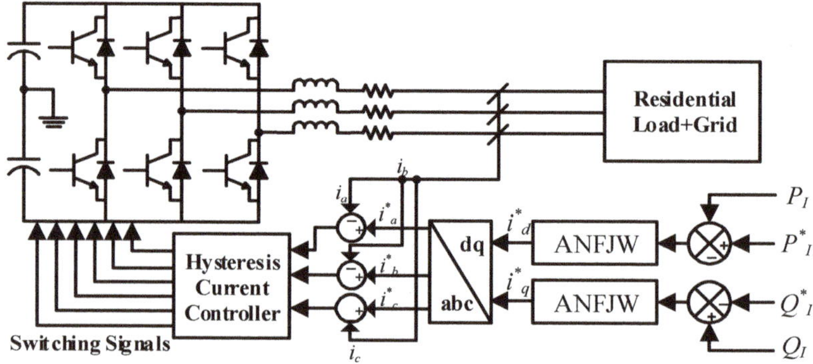

Figure 5. Working of the proposed strategy.

Each controller was modelled in seven layers, as shown in Figure 6. The first three layers formed the antecedent part, and the next four layers formed the consequent part of the ANFJW controller. The number of inputs in the first layer was equivalent to the $n\prime$ number of nodes, which were used for further distribution as inputs.

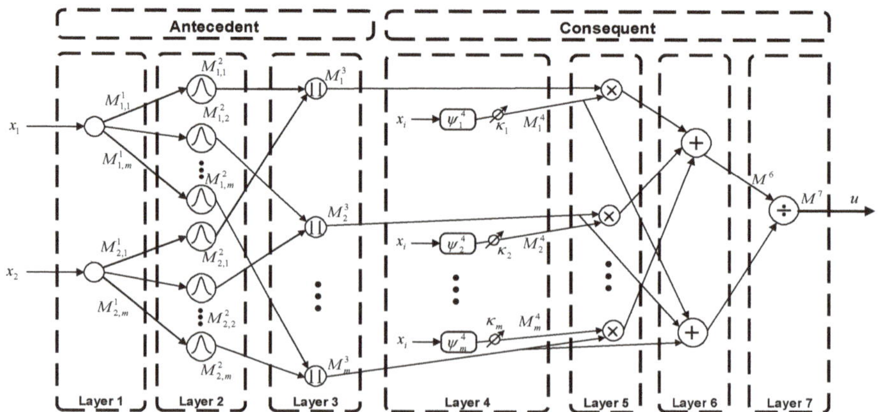

Figure 6. Structure of the adaptive neuro-fuzzy Jacobi wavelet (ANFJW) controller.

N_i^k and M_i^k describe the input and output of a node in *kth* layer, respectively.

First Layer:

It combines the numbers of the input variables, and are then transferred to the next layer by the nodes from the first layer, which is written as follows:

$$N_i^1(k) = x_i^1(k) \tag{2}$$

$$M_i^1(k) = N_i^1(k) = x_i^1(k) \tag{3}$$

where $i = 1, 2, \ldots, m$ and k stands for the number of nodes and iteration, respectively.

Second Layer:

In this layer, the fuzzy system receives the linguistic terms and their degree of membership of each input. The Gaussian membership function (GMF) is used for computing the linguistic terms of each input.

$$N_i^2(k) = M_i^1(k) = x_i^1(k) \tag{4}$$

$$N_i^2(k) = \mu_i^2 = e^{-0.5\left(\frac{N_i^2(k)-m_i}{\sigma_i}\right)^2} = e^{-0.5\left(\frac{x_i^1(k)-m_{ij}^2}{\sigma_{ij}^2}\right)^2} \tag{5}$$

where m_{ij} and σ_{ij} denote the center and variance of GMF, respectively.

Third Layer:

In this layer, the product of membership function is performed, where the M_{in} operator is used to find the output value.

$$N_i^3(k) = M_i^2 = \mu_i^2 = e^{-0.5\left(\frac{x_i^1(k)-m_{ij}^2}{\sigma_{ij}^2}\right)^2} \tag{6}$$

$$M_i^3(k) = \mu_i^3 = \prod_{i=1}^{m} \mu_i^2 = \prod_{i=1}^{m} e^{-0.5\left(\frac{x_i^1(k)-m_{ij}^2}{\sigma_{ij}^2}\right)^2} \tag{7}$$

Fourth Layer:

The Jacobi wavelet function is used in this layer, which is written as follows:

$$\psi_{nq}^{\alpha,\beta}(x) = \begin{cases} 2^{0.5h}\lambda_q^{\alpha,\beta}(2^h x - 2n + 1), \ \forall \ \frac{n-1}{2^{h-1}} \le x < \frac{n}{2^{h-1}} \\ 0, \qquad\qquad o.w \end{cases} \tag{8}$$

where

$$\lambda_q^{\alpha,\beta} = \sqrt{\frac{(2q+\alpha+\beta+1)\Gamma(2q+\alpha+\beta+1)m!}{2^{\alpha+\beta+1}\Gamma(q+\alpha+1)\Gamma(q+\beta+1)}} \tag{9}$$

$$J_q^{\alpha+\beta} = \sum_{i=0}^{q} \binom{q+\alpha}{i}\binom{q+\beta}{q-i}\left(\frac{k-1}{2}\right)^{q-i}\left(\frac{k+1}{2}\right)^{i}, \quad k \in [-1,1] \tag{10}$$

here,$x = N_i^4$, where the input of this layer is $N_i^4(k) = \psi_{nq}^{\alpha+\beta}(N_i^4)$, and the output of this layer is written as follows:

$$M_i^4 = f(x) = \sum_{n=0}^{\infty}\sum_{q\in z} \kappa_{nq}\psi_{nq}^{\alpha+\beta}(x) \tag{11}$$

The main objective is to reduce the error between the reference power and desire power. The proposed controller works on the error, e, which is the difference between the reference power and desire power, as illustrated in Figure 7. The proposed control strategy processes the error, and it is written as follows:

$$e = \left[e_p = P_I^* - P_I \ \ \& \ e_q = Q_I^* - Q_I\right] \quad (12)$$

Fifth Layer:

In this layer, the output of the antecedent and consequents parts is multiplied, and then added for each input.

$$M_i^5(k) = \sum_{i=1}^{m} \beta_i^4 \mu_i^3 \quad (13)$$

Sixth Layer:

The summation of rules (third layer output) is performed in this layer.

$$M_i^6(k) = \sum_{i=1}^{m} \mu_i^3 \quad (14)$$

Seven Layer:

In this layer, the output of ANFJW is calculated as follows:

$$u_{inv} = M_i^7(k) = \frac{M_i^5(k)}{M_i^6(k)} = \frac{\sum_{i=1}^{m} \beta_i^4 \mu_i^3}{\sum_{i=1}^{m} \mu_i^3} \quad (15)$$

This output is used in the duty cycle to generate the control commands to the switches of the inverter.

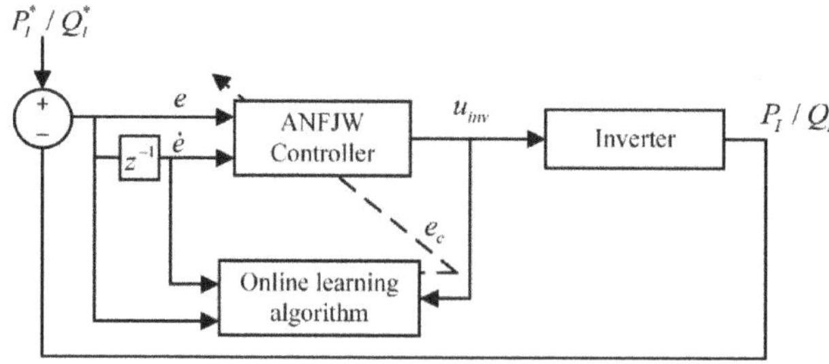

Figure 7. Working of the proposed strategy.

Below, it is explained how the parameters of ANFJW controller are updated. The output of the proposed controller is given as follows:

$$E = \frac{1}{2}(y_d - y)^2 + \frac{l}{2}u_{inv}^2 \quad (16)$$

and the objective function to be minimized is given as follows:

$$E = \frac{1}{2}(y_d - y)^2 + \frac{l}{2}u_{inv}^2 \quad (17)$$

The parameters to be updated are the following:

$$\xi = [m, \sigma, c]^T$$

These parameters are updated online through the gradient back propagation method. The gradient-based update equations are given as follows:

$$\xi(k+1) = \xi(k) + \lambda \frac{\partial E}{\partial \xi} \tag{18}$$

$$\xi(k+1) = \xi(k) + \lambda \frac{\partial}{\partial \xi}\left(\frac{1}{2}(y_d - y)^2 + \frac{l u_{inv}^2}{2}\right) \tag{19}$$

$$\xi(k+1) = \xi(k) + \lambda\left[-e \frac{\partial y}{\partial u_{inv}} \frac{\partial u_{inv}}{\partial \xi} + l u_{inv}\right] \frac{\partial u_{inv}}{\partial \xi} \tag{20}$$

where $e = y_d - y$

$$\xi(k+1) = \xi(k) + \lambda\left[-e \frac{\partial y}{\partial u_{inv}} + l u_{inv}\right] \frac{\partial u_{inv}}{\partial \xi} \tag{21}$$

$$\xi(k+1) = \xi(k) + \lambda\left[-\left(e \frac{\partial y}{\partial u_{inv}} - l u_{inv}\right)\right] \frac{\partial u_{inv}}{\partial \xi} \tag{22}$$

$$\xi(k+1) = \xi(k) + \lambda \aleph \frac{\partial u_{inv}}{\partial \xi} \tag{23}$$

$$\aleph = -\left(e \frac{\partial y}{\partial u_{inv}} - l u_{inv}\right) = -(y_d - y) \frac{\partial y}{\partial u_{inv}} - l u_{inv}) \tag{24}$$

Now, for simplification, $\frac{\partial y}{\partial u_{inv}} = 1$, and $\frac{\partial u_{inv}}{\partial \xi}$ is evaluated using the chain rule. The updated equations are written as follows:

$$c_{ij}(k+1) = c_{ij}(k) + \lambda \aleph \frac{\partial u_{inv}}{\partial c_{ij}} \tag{25}$$

$$\sigma_{ij}(k+1) = \sigma_{ij}(k) + \lambda \aleph \frac{\partial u_{inv}}{\partial \sigma_{ij}} \tag{26}$$

$$m_{ij}(k+1) = m_{ij}(k) + \lambda \aleph \frac{\partial u_{inv}}{\partial m_{ij}} \tag{27}$$

By calculating the partial derivative for the individual parameters, it is obtained as follows:

$$\frac{\partial u_{inv}}{\partial c_{ij}} = \frac{\partial u_{inv}}{\partial X_i^4} \frac{\partial X_i^4}{\partial c_{ij}} \tag{28}$$

$$\frac{\partial u_{inv}}{\partial c_{ij}} = \frac{Y_i^2}{Y_i^6} X_i^4 \tag{29}$$

$$\begin{aligned} \frac{\partial u_{inv}}{\partial \sigma_{ij}} &= \frac{\partial u_{inv}}{\partial Y_i^3} \frac{\partial Y_i^3}{\partial \sigma_{ij}} = \left[-\frac{Y_i^3 u_{INV}}{Y_i^6} + \frac{X_i^4}{Y_i^6}\right] Y_i^3 \cdot 2 \frac{(k_i - \sigma_{ij})}{m_{ij}^2} \\ &= \frac{Y_i^6 \frac{\partial}{\partial Y_i^3}(Y_i^5) - (Y_i^5) \frac{\partial}{\partial Y_i^3}(Y_i^6)}{(Y_i^6)^2} \frac{\partial Y_i^3}{\partial \sigma_{ij}} \\ &= \frac{X_i^4 Y_i^6 - Y_i^5 \cdot 1}{(Y_i^6)^2} \frac{\partial Y_i^3}{\partial \sigma_{ij}} = \frac{X_i^4}{Y_i^6} - \frac{Y_i^5}{Y_i^6} \cdot \frac{1}{Y_i^6} \cdot \frac{\partial Y_i^3}{\partial \sigma_{ij}} \\ &= \left[\frac{Y_i^2 - u_{inv}}{Y_i^6}\right] \frac{\partial Y_i^3}{\partial \sigma_{ij}} \end{aligned} \tag{30}$$

where

$$Y_i^3(k) = \mu_i^3 = \prod_{i=1}^{m} \mu_i^2 = \prod_{i=1}^{m} e^{-\frac{1}{2}\left(\frac{k_i - m_{ij}^2}{\sigma_{ij}^2}\right)^2} \tag{31}$$

So, Equation (31) becomes:

$$= \left[\frac{X_i^4 - u_{inv}}{\mu_i^3}\right] \cdot \prod_{i=1, i\neq j}^{m} \mu_i^2 \cdot \frac{\partial \mu_i^2}{\partial \sigma_{ij}} \tag{32}$$

$$\frac{\partial u_{inv}}{\partial \sigma_{ij}} = \left[\frac{X_i^4 - u_{inv}}{\mu_i^3}\right] \cdot \prod_{i=1, i\neq j}^{m} \mu_i^2 \left[\mu_i^2 \frac{2(k_i - \sigma_{ij})}{m_{ij}^2}\right] \tag{33}$$

Similarly,

$$\begin{aligned} \frac{\partial u_{INV}}{\partial m_{ij}} &= \frac{\partial u_{INV}}{\partial \mu_i^3} \frac{\partial \mu_i^3}{\partial m_{ij}} = \left[\frac{X_i^4 - u_{inv}}{\mu_i^3}\right] \frac{\partial \mu_i^3}{\partial m_{ij}} \\ &= \left[\frac{X_i^4 - u_{inv}}{\mu_i^3}\right] \prod_{i=1, i\neq j}^{m} \mu_i^2 \left[\mu_i^2 \frac{2(k_i - \sigma_{ij})^2}{m_{ij}^3}\right] \\ &= \left[\frac{X_i^4 - u_{inv}}{\mu_i^3}\right] \mu_i^3 \cdot \frac{2(k_i - \sigma_{ij})^2}{m_{ij}^3} \end{aligned} \tag{34}$$

Using the values of Equations (20), (33), and (34) in Equations (25)–(27), the following final updated equations are obtained.

$$c_{ij}(k+1) = c_{ij}(k) + \lambda \aleph \frac{Y_i^2}{Y_i^6} X_i^4 \tag{35}$$

$$\sigma_{ij}(k+1) = \sigma_{ij}(k) + \lambda \aleph \left[\frac{X_i^4 - u_{inv}}{\mu_i^3}\right] \cdot \prod_{i=1, i\neq j}^{m} \mu_i^2 \left[\mu_i^2 \frac{2(k_i - \sigma_{ij})}{m_{ij}^2}\right] \tag{36}$$

$$m_{ij}(k+1) = m_{ij}(k) + \lambda \aleph \left[\frac{X_i^4 - u_{inv}}{\mu_i^3}\right] \mu_i^3 \cdot \frac{2(k_i - \sigma_{ij})^2}{m_{ij}^3} \tag{37}$$

3. Energy Management and Supervisory Control System

A supervisory control approach was designed to provide the required power demand during the day and after sunset by using the proposed algorithm, as illustrated in Figure 8.

The proposed supervisory control system controls the subsystems (PV, UC, battery, and power converters), as well as the whole microgrid. According to the implemented algorithm, the net load demand must be satisfied from the generation of the PV array. If PV cannot satisfy the net demand, then battery bank will supply the remaining power, if its charge level is sufficient, i.e., above 20% (state 1). If the PV and battery bank cannot provide the total demand, then the remaining power will be taken from the SC if its charge level is above 20% (state 2), followed by the utility grid (state 3). Similarly, if PV generates more power than the demand, the remaining power will be used to charge the battery and then the SC, if their states of charges are below 90% (state 4 and state 5). If the battery and SC are charged, then the remaining power will be transferred to the public grid (state 6).

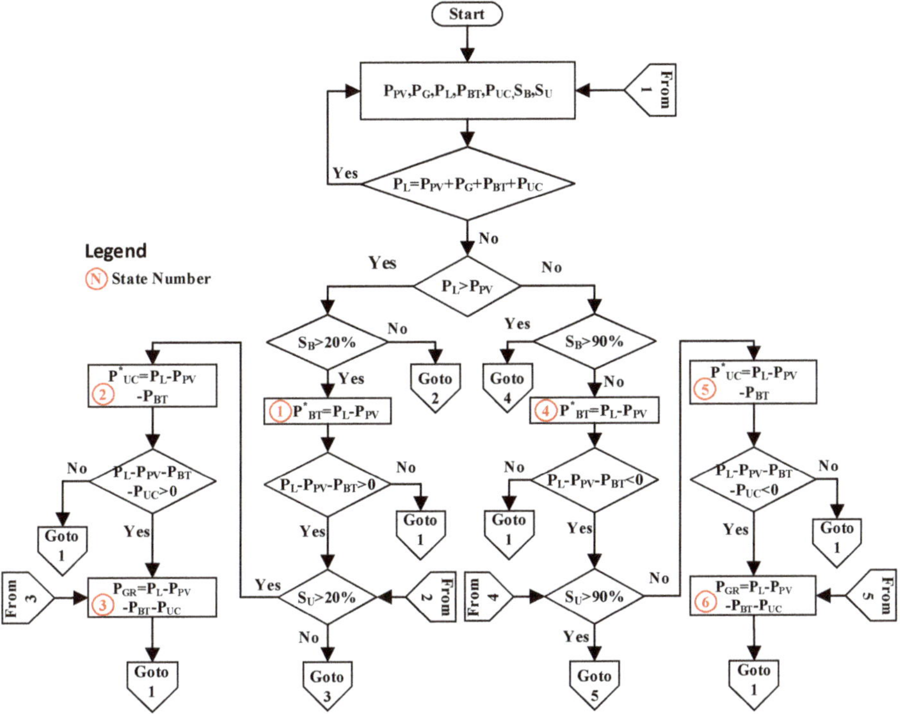

Figure 8. Flowchart of the energy management and supervisory control system (EMSCS).

The EMSCS layer power provides the required control signals to the power converters connected to the inputs/outputs of the components used in the microgrid. The descriptions of the parameters are defined in Table 1.

Table 1. Inputs/outputs of the proposed EMSCS.

Symbol	Description
P_L	Local Load Power
P_G	Grid Power
P_B	Battery Power
P_U	Ultra-capacitor Power
P_{PV}	PV Power
S_B	SoC of Battery
S_U	SoC of UC
P_{BDR}	Discharging Reference Power of battery
P_{BCR}	Charging Reference Power of Battery
P_{UDR}	UC Discharging Reference Power
U_{UCR}	UC Charging Reference Power
P_{GR}	Grid Reference Power

The working of the algorithm is discussed as below.

1. All of the control signals are generated, i.e., P_{PV}, P_G, P_L, P_B, P_U, S_B, and S_U.
2. Check $P_L = P_{PV} \pm P_G \pm P_U \pm P_B$, go to 1 if this condition is true, and if not then follow next step.
3. Check $P_L > P_{PV}$, if it is true, go to step 9, and if not then check the next condition.
4. Check $S_B > 20\%$, if it is true, then discharge the battery, and go to next step, otherwise go to step 2.

5. Check the condition $P_L - P_{PV} - P_B > 0$, if this is true, then go to the next step, otherwise go to step 6.
6. Check $S_U > 20\%$, if it is true, then discharge the UC, and go to the next step, otherwise go to step 8.
7. Check the condition $P_L - P_{PV} - P_B - P_U > 0$, if it is true, then go to the next step, otherwise go to step 1.
8. Using all of the remaining deficient power reference to the grid and go to step 1.
9. Check $S_B > 90\%$, if it is not true, then charge the battery and go to the next step, otherwise go to step 11.
10. Check the condition $P_L - P_{PV} - P_B < 0$, if true, then go to the next step, otherwise go to step 1.
11. Check $S_U > 90\%$, if it is not true, then charge the UC and go to the next step, otherwise go to step 13.
12. Check the condition $P_L - P_{PV} - P_B - P_U < 0$, if true, then go to the next step, otherwise step 1.
13. Provide all of the net surplus power to the utility grid and go to step 1.

4. Simulations

The proposed microgrid was simulated for a complete full day under real weather patterns, i.e., ambient temperature (°C) and solar irradiance (W/m^2), taken at Islamabad, Pakistan. Both parameters were recorded on an hourly basis, as presented in Figure 9. The intensity of irradiance fluctuated during the day, depending on the sunrise and sunset. From Figure 9, the sun appeared at 07:00 a.m. and set at 17:20 p.m. The average solar irradiance during the daytime was 990 (W/m^2). Likewise, the average temperature during the daytime reached 40 °C, while at night, it went down to 19 °C.

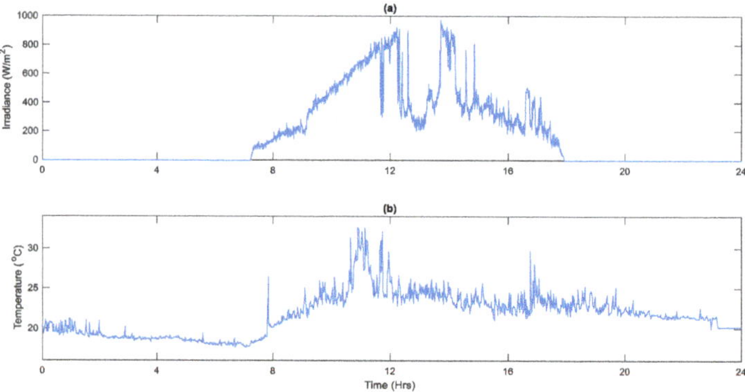

Figure 9. Weather data: (**a**) irradiance and (**b**) temperature.

Figure 10 illustrates the power generated by the PV (denoted as violet), battery (denoted as red) main inverter (denoted as dark goldenrod), UC (denoted as light green), utility grid (denoted as black), and total demand (denoted as light blue). In order to explain Figure 10, it is divided into six states, as indicated in the flow chart (Figure 8). From t = 00:00 to 07:10 a.m., the PV output power was zero because of the nonappearance of sun irradiance. The overall demand was applied on the battery system. During this interval, the proposed EMSCS checked the state of charge (SoC) of the battery, and as its SoC was greater than 20%, the battery provided the required net power (around 150 kW). The battery SoC was reduced to 68%, as shown in Figure 11. The EMSCS was operating in state 1. At t = 07:10 to 09:00 a.m., the PV system started producing power, but it was still not enough to overcome the demand. Meanwhile, the output power of the PV started to increase. At the end of the interval, the PV and battery provided 90 kW each with the battery SoC at 62%. The EMSCS was still operating in state 1. At t = 09:00 to 09:48 a.m., because of the slow response time of the battery, the cumulative power supply (battery (52 kW) + PV system (179 kW)) exceeded the power demand

(198 kW). Therefore, the UC started charging using 33 kW of power when the SoC went from to 84% from 73%, and the EMSCS was shifted to state 5.

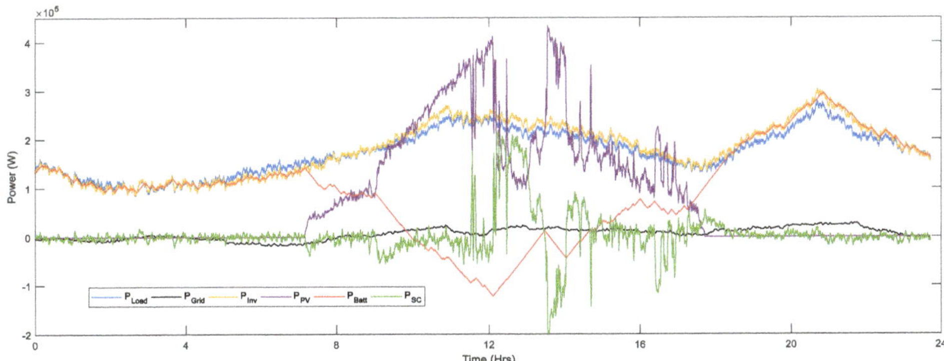

Figure 10. Power exchange and sharing among energy sources.

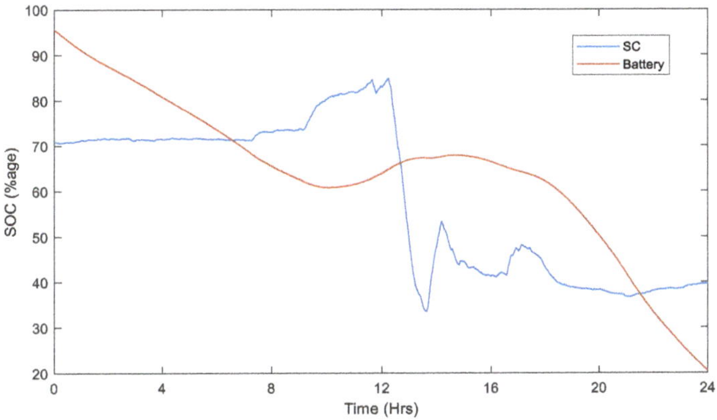

Figure 11. States of charge (SoCs) of the ultra-capacitor (UC) and battery.

At t = 09:48 to 11:30 a.m., the PV not only provided sufficient power to the load, but also charged the battery and UC. In this interval, the EMSCS was in states 4 and 5. At t = 11:30 to 11:39 a.m., 12:10 to 13:27 p.m., and 14:03 to 14:40 p.m., the PV output power suddenly dropped due to partially clouded conditions. In this crucial time, the UC assisted the microgrid and fulfilled the power gap with slight support from the grid station. In this interval, the EMSCS rapidly shifted to states 2 and 3. At t = 14:40 to 17:36 p.m., the PV output power started decreasing because of the evening time. The battery could fulfil the power deficiency and kept the EMSCS in state 1. At t =17:36 to 18:15 p.m., the battery, UC, and grid station provided backup to the microgrid because of the deficient power. The EMSCS shifted between states 1, 2, and 3. Whereas at t= 18:15 to 24:00 p.m., with no PV power and also because of the non-peak hours, the grid and battery fulfilled the load demand, and the battery SoC dropped to 21%. In this interval, the EMSCS moved between states 1 and 3. The operating states for 24 h are shown in Figure 12. In this manner, the power exchange and sharing among PV, battery, UC, utility grid, and domestic load happened for 24 h under EMSCS.

For better power quality and power flow from the DC link to the rest of the system, the inverter was controlled via ANFJW controllers. The results of the proposed ANFJW controllers were compared with the existing intelligent controllers, such as NFC and fuzzy logic controller (FLC), as well as the classical controller (i.e., PID), using the same weather and load conditions. Figure 13 shows the performance of the inverter (active power) for the different controllers in terms of power transfer. The load condition had two peaks of 275 kW and 307 kW at t = 11:00 a.m. and 21:00 p.m., respectively. The zoomed figures show the performance of the difference controllers in detail. It can be clearly noticed from the zoomed figures (Figure 13) that the proposed controller, ANFJW (denoted as red), and reference signal (donated as dashed-blue) superimposed each other, while the other controllers also tried to track the reference signal, but were unable to because of overshoots. Similarly, Figure 14 gives the performance of the inverter in terms of the reactive power transfer. For instance, from the zoomed portion (i.e., at 04:45 to 04:48 a.m.) in Figure 14, the proposed controller tracked the reference power very quickly, while the other controllers took some time with the overshoots. The same performance can also be seen at 19:57 to 20:03 p.m. in the zoomed portion. Hence, the proposed ANFJW controllers show a better performance at different intervals for both real and reactive power transfer.

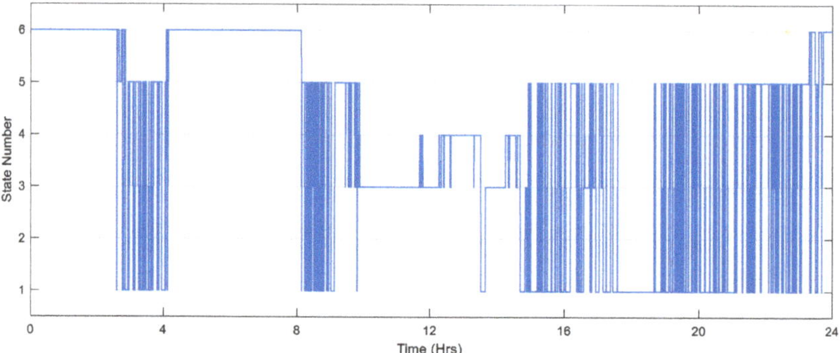

Figure 12. Operating states.

Furthermore, the power quality analyses in terms of the THD and frequency of all of controllers was performed and are illustrated in Figures 15 and 16, respectively. According to IEEE standards, the standard limits for frequency and THD are ±0.8% and 5%, respectively. In this work, the operating load voltage RMS and frequency were 440 V and 50 Hz, respectively. The THD shown in Figure 15 is small with the proposed controller, which is around 2.37%. Similarly, the change in frequency was below 0.02%, as it can be seen at different intervals in the zoomed windows in Figure 16.

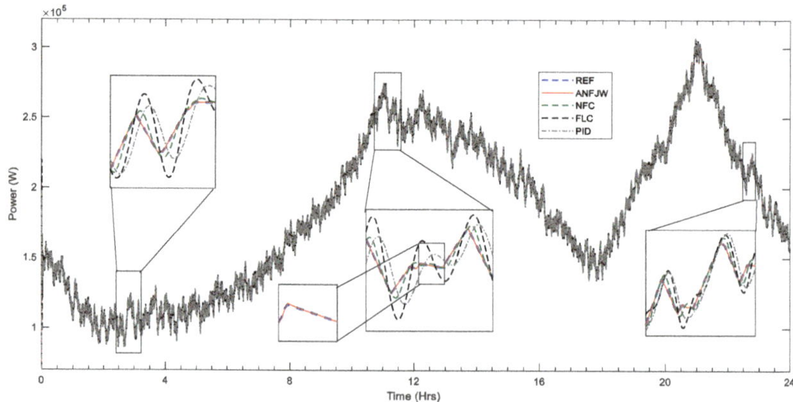

Figure 13. Output real power comparison at different intervals.

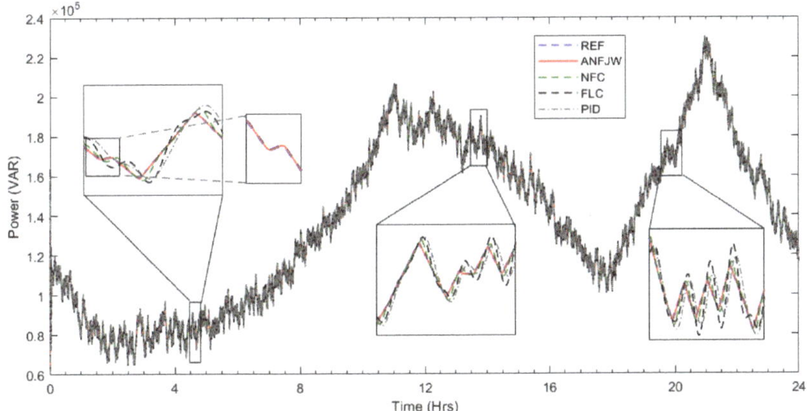

Figure 14. Output reactive power comparison at different intervals.

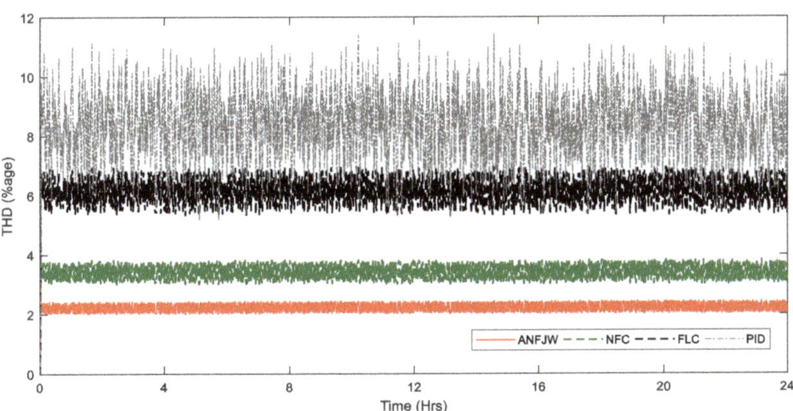

Figure 15. Total harmonic distortion (THD) comparison.

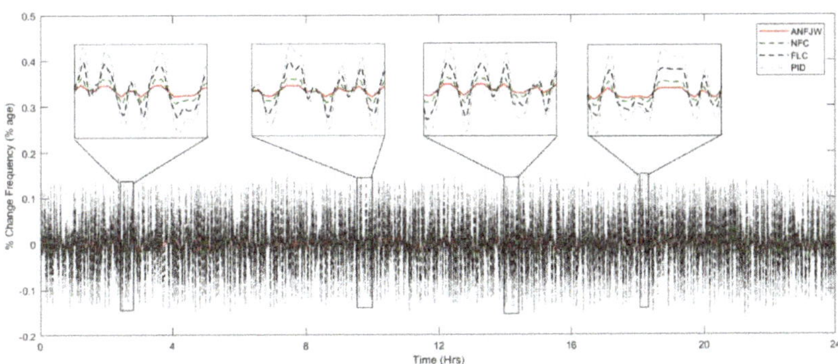

Figure 16. Comparison of change in frequency.

To show further better the performance, the efficiency of the inverter for all controllers using Equation (1) was calculated for 24 h and is shown in Figure 17. The real and reactive power transfer efficiencies of the inverter with the ANFJW controllers was greater, i.e., 99.05% and 99.08%, followed by NFC, FLC, and then PID.

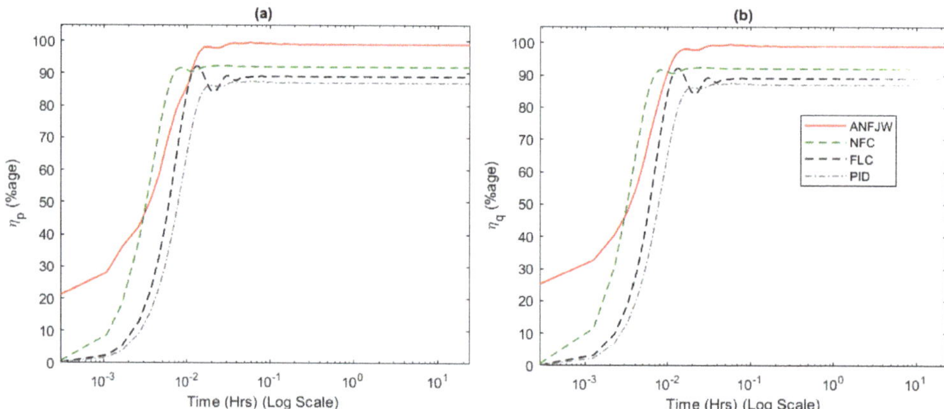

Figure 17. Power efficiency comparison of different controllers: (**a**) real power and (**b**) reactive power.

For further analysis, the dynamic performance of all of the controllers through the index parameters, i.e., integral absolute error (IAE), integral square error (ISE), integral time absolute error (ITAE), integral time square error (ITSE), and mean relative error (MRE), under the same operating conditions, were calculated using Equations (38)–(41) for active power, and using Equations (42)–(45) for reactive power, and are illustrated in Figures 18 and 19, respectively. All of the index parameters were much smaller in the case of the proposed controller, which showed a superior dynamic performance. All of the comparisons are summarized in Table 2.

$$IAE_p = \int_0^t |e_p(t)| dt \tag{38}$$

$$ISE_p = \int_0^t e_p^2(t) dt \tag{39}$$

$$ITAE_p = \int_0^t t|e_p(t)| dt \tag{40}$$

$$ITSE_p = \int_0^t te_p^2(t) dt \tag{41}$$

$$IAE_q = \int_0^t |e_q(t)| dt \tag{42}$$

$$ISE_q = \int_0^t e_q^2(t) dt \tag{43}$$

$$ITAE_q = \int_0^t t|e_q(t)| dt \tag{44}$$

$$ITSE_q = \int_0^t te_q^2(t) dt \tag{45}$$

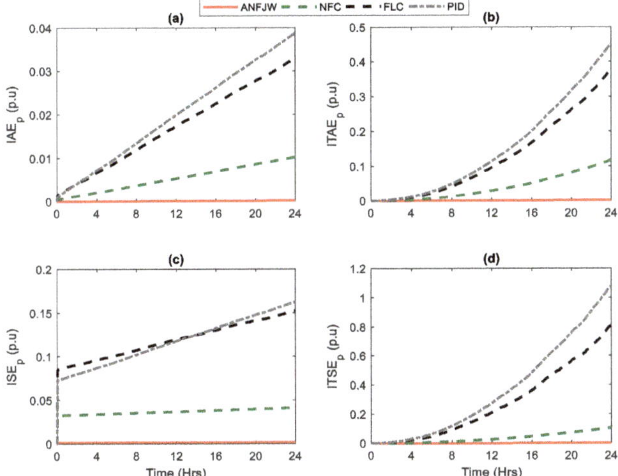

Figure 18. Dynamic performance comparison using active power: (**a**) integral absolute error (IAE), (**b**) integral time absolute error (ITAE), (**c**) integral square error (ISE), and (**d**) integral time square error (ITSE).

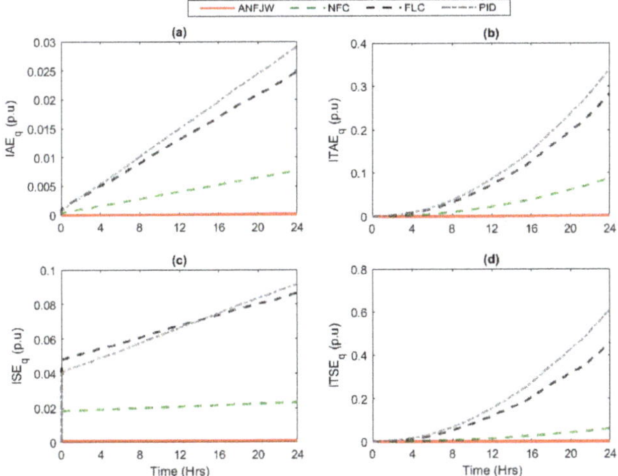

Figure 19. Dynamic performance comparison using reactive power: (**a**) IAE, (**b**) ITAE, (**c**) ISE, and (**d**) ITSE.

Table 2. Comparison of efficiencies and dynamic response.

Controllers	Output Power	η_{IN} (% Age)	THD (% Age)	IAE (p.u)	ITAE (p.u)	ISE (p.u)	ITSE (p.u)
ANFJW	Active	99.05	2.37	0.00017	0.00166	0.00166	0.00052
	Reactive	99.08		0.00012	0.00128	0.00093	0.00003
NFC	Active	92.17	3.63	0.0102	0.1169	0.0413	0.1052
	Reactive	92.25		0.0076	0.0879	0.0233	0.0597
FLC	Active	89.11	6.54	0.0329	0.3758	0.1526	0.8016
	Reactive	89.18		0.0247	0.2829	0.0864	0.4573
PID	Active	86.94	8.96	0.0386	0.4526	0.1619	1.078
	Reactive	87.04		0.0290	0.3394	0.0917	0.6097

5. Conclusions

This paper presented a novel improved adaptive NFC of inverter and supervisory energy management of a microgrid. The improvement in the existing NFC was performed by the integration of the Jacobi wavelet. Because of the excellent time-localized behavior of the Jacobi wavelet, the proposed controller did not stop in the first local minima, but it searched for the optimal minima. This yielded excellent results in terms of power transfer, inverter output efficiency, and load voltage frequency. The EMSCS controlled the individual components as well as the entire system to ensure the following: (1) maximize output power, (2) appropriate sharing of UC/battery power and energy sources, and (3) continuity of power for a 24 h power supply with reliability and lessening the differences in output power under unfavorable weather conditions and inadequate storage.

Author Contributions: Conceptualization, T.K. and S.Z.H.; methodology, T.K., M.K., and S.Z.H.; software, T.K.; validation, T.K. and S.Z.H.; formal analysis, M.K., V.S.P., and L.M.F.-R.; investigation, T.K. and M.K.; resources, S.Z.H.; data curation, S.Z.H.; writing (original draft preparation), T.K., M.K., and S.Z.H.; writing (review and editing), V.S.P. and L.M.F.-R.; visualization, T.K. and S.Z.H.; supervision, M.K., V.S.P., and L.M.F.-R.; project administration, M.K., V.S.P., and L.M.F.-R.; funding acquisition, M.K., V.S.P., and L.M.F.-R. All authors have read and agreed to the published version of the manuscript.

Funding: The work of Murat Karabacak and Tariq Kamal is sponsored by the Scientific and Technological Research Council of Turkey under project number 5190011. The work of Vedran S. Perić is supported by the Bavarian Government and Deutsche Forschungsgemeinschaft (DFG) under project number 350746631. The work of Luis M. Fernández-Ramírez and Tariq Kamal is sponsored by the Spanish Ministry of Science, Innovation, and Universities under project number RTI2018-095720-B-C32.

Acknowledgments: Authors are thankful to all organizations stated in the funding portion.

Conflicts of Interest: There is no conflict of interest.

Abbreviations

ANFJW	Adaptive neuro-fuzzy Jacobi wavelet
EMSCS	Energy management and supervisory control system
FLC	Fuzzy logic controller
GMF	Gaussian membership function
IAE	Integral absolute error
IC	Incremental conductance
ISE	Integral square error
ITAE	Integral time absolute error
ITSE	Integral time square error
MPPT	Maximum power point tracking
MRE	Mean relative error
NF	Neuro-fuzzy
NN	Neural network
PV	Photovoltaic
RES	Renewable energy sources
THD	Total harmonic distortion
UC	Ultra-capacitor

References

1. Ozdemir, S.; Kaplan, O.; Sefa, I.; Altin, N. Fuzzy PI controlled inverter for grid interactive renewable energy systems. *IET Renew. Power Gener.* **2015**, *9*, 729–738.
2. Hossain, M.; Pota, H.; Issa, W.; Hossain, M.; Hossain, M.A.; Pota, H.R.; Issa, W.; Hossain, M.J. Overview of AC Microgrid Controls with Inverter-Interfaced Generations. *Energies* **2017**, *10*, 1300. [CrossRef]
3. Altin, N.; Ozdemir, S. Three-phase three-level grid interactive inverter with fuzzy logic based maximum power point tracking controller. *Energy Convers. Manag.* **2013**, *69*, 17–26. [CrossRef]

4. Alsayari, N.; Chilipi, R.; Al Hosani, K.; Almaskari, F. Grid synchronization and control of distributed generation unit with flexible load compensation capabilities using multi-output LMS-filter. *Int. J. Electr. Power Energy Syst.* **2017**, *93*, 253–265. [CrossRef]
5. Al Sayari, N.; Chilipi, R.; Barara, M. An adaptive control algorithm for grid-interfacing inverters in renewable energy based distributed generation systems. *Energy Convers. Manag.* **2016**, *111*, 443–452. [CrossRef]
6. Cecati, C.; Dell'Aquila, A.; Liserre, M.; Monopoli, V.G. Design of H-bridge multilevel active rectifier for traction systems. *IEEE Trans. Ind. Appl.* **2003**, *39*, 1541–1550. [CrossRef]
7. Kaźmierkowski, M.P.; Krishnan, R.; Blaabjerg, F. *Control in Power Electronics: Selected Problems*; Academic Press: San Diego, CA, USA, 2002; ISBN 9780080490786.
8. Yuan, X.; Merk, W.; Stemmler, H.; Allmeling, J. Stationary-frame generalized integrators for current control of active power filters with zero steady-state error for current harmonics of concern under unbalanced and distorted operating conditions. *IEEE Trans. Ind. Appl.* **2002**, *38*, 523–532. [CrossRef]
9. Ren, X.; Lyu, Z.; Li, D.; Zhang, Z.; Zhang, S. Synchronization signal extraction method based on enhanced DSSOGI-FLL in power grid distortion. *Syst. Sci. Control Eng.* **2018**, *6*, 305–313. [CrossRef]
10. Li, S.; Haskew, T.A.; Hong, Y.K.; Xu, L. Direct-current vector control of three-phase grid-connected rectifier-inverter. *Electr. Power Syst. Res.* **2011**, *81*, 357–366. [CrossRef]
11. Sędłak, M.; Styński, S.; Kaźmierkowski, M.P.; Malinowski, M. Three-level four-leg flying capacitor converter for renewable energy sources. *Przeglad Elektrotechniczny* **2012**, *88*, 6–11.
12. Chandran, B.P.; Selvakumar, A.I.; Mathew, F.M. Integrating multilevel converters application on renewable energy sources—A survey. *J. Renew. Sustain. Energy* **2018**, *10*, 065502. [CrossRef]
13. Verdugo, C.; Kouro, S.; Rojas, C.A.; Perez, M.A.; Meynard, T.; Malinowski, M. Five-Level T-type Cascade Converter for Rooftop Grid-Connected Photovoltaic Systems. *Energies* **2019**, *12*, 1743. [CrossRef]
14. Mishra, N.; Singh, B. Solar PV Grid Interfaced System with Neutral Point Clamped Converter for Power Quality Improvement. *J. Inst. Eng. Ser. B* **2018**, *99*, 605–612. [CrossRef]
15. Carrasco, J.M.; Franquelo, L.G.; Bialasiewicz, J.T.; Galvan, E.; PortilloGuisado, R.C.; Prats, M.A.M.; Leon, J.I.; Moreno-Alfonso, N. Power-Electronic Systems for the Grid Integration of Renewable Energy Sources: A Survey. *IEEE Trans. Ind. Electron.* **2006**, *53*, 1002–1016. [CrossRef]
16. Shadmand, M.B.; Li, X.; Balog, R.S.; Rub, H.A. Model predictive control of grid-tied photovoltaic systems: Maximum power point tracking and decoupled power control. In Proceedings of the 2015 First Workshop on Smart Grid and Renewable Energy (SGRE), Doha, Qatar, 22–23 March 2015; pp. 1–6.
17. Hu, J.; Zhu, J.; Dorrell, D.G. Model-predictive control of grid-connected inverters for PV systems with flexible power regulation and switching frequency reduction. *IEEE Trans. Ind. Appl.* **2015**, *51*, 587–594. [CrossRef]
18. Cecati, C.; Ciancetta, F.; Siano, P. A Multilevel Inverter for Photovoltaic Systems with Fuzzy Logic Control. *IEEE Trans. Ind. Electron.* **2010**, *57*, 4115–4125. [CrossRef]
19. Hannan, M.A.; Ghani, Z.A.; Hoque, M.M.; Ker, P.J.; Hussain, A.; Mohamed, A. Fuzzy Logic Inverter Controller in Photovoltaic Applications: Issues and Recommendations. *IEEE Access* **2019**, *7*, 24934–24955. [CrossRef]
20. Mao, Y.; Yang, Y. Backstepping sliding mode control of grid-connected inverters. *Int. J. Electron. Lett.* **2017**, *5*, 314–326. [CrossRef]
21. Sun, Y.; Li, S.; Lin, B.; Fu, X.; Ramezani, M.; Jaithwa, I. Artificial Neural Network for Control and Grid Integration of Residential Solar Photovoltaic Systems. *IEEE Trans. Sustain. Energy* **2017**, *8*, 1484–1495. [CrossRef]
22. Altin, N.; Sefa, İ. dSPACE based adaptive neuro-fuzzy controller of grid interactive inverter. *Energy Convers. Manag.* **2012**, *56*, 130–139. [CrossRef]
23. Vazquez, S.; Leon, J.I.; Franquelo, L.G.; Carrasco, J.M.; Martinez, O.; Rodriguez, J.; Cortes, P.; Kouro, S. Model Predictive Control with constant switching frequency using a Discrete Space Vector Modulation with virtual state vectors. In Proceedings of the 2009 IEEE International Conference on Industrial Technology, Gippsland, VIC, Australia, 10–13 February 2009; pp. 1–6.
24. Lee, H.; Utkin, V.I. Chattering suppression methods in sliding mode control systems. *Annu. Rev. Control* **2007**, *31*, 179–188. [CrossRef]
25. Liu, Y.-H.; Liu, C.-L.; Huang, J.-W.; Chen, J.-H. Neural-network-based maximum power point tracking methods for photovoltaic systems operating under fast changing environments. *Sol. Energy* **2013**, *89*, 42–53. [CrossRef]

26. Badar, R.; Khan, L. Adaptive Neuro Fuzzy Wavelet Based SSSC Damping Control Paradigm. In Proceedings of the 2012 10th International Conference on Frontiers of Information Technology, Islamabad, Pakistan, 17–19 December 2012; pp. 101–106.
27. Kamal, T.; Karabacak, M.; Hassan, S.Z.; Fernández-Ramírez, L.M.; Roasto, I.; Khan, L. An indirect adaptive control paradigm for wind generation systems. In *Advanced Control and Optimization Paradigms for Wind Energy Systems*; Springer: Singapore, 2019; pp. 235–257.
28. Hassan, S.; Li, H.; Kamal, T.; Arifoğlu, U.; Mumtaz, S.; Khan, L. Neuro-Fuzzy Wavelet Based Adaptive MPPT Algorithm for Photovoltaic Systems. *Energies* **2017**, *10*, 394. [CrossRef]
29. Banakar, A.; Azeem, M.F. Artificial wavelet neural network and its application in neuro-fuzzy models. *Appl. Soft Comput. J.* **2008**, *8*, 1463–1485. [CrossRef]
30. Yusuf, O.; Yilmaz, S. An Adaptive Fuzzy Wavelet Network with Gradient Learning for Nonlinear Function Approximation. *J. Intell. Syst.* **2014**, *23*, 201–212.
31. Hassan, S.Z.; Kamal, T.; Mumtaz, S.; Khan, L. An online self recurrent direct adaptive neuro-fuzzy wavelet based control of photovoltaic systems. In *Solar Photovoltaic Power Plants. Advanced Control and Optimization Techniques*; Springer: Singapore, 2019; pp. 233–250.
32. Kumar, S.A.; Subathra, M.S.P.; Kumar, N.M.; Malvoni, M.; Sairamya, N.J.; George, S.T.; Suviseshamuthu, E.S.; Chopra, S.S. A Novel Islanding Detection Technique for a Resilient Photovoltaic-Based Distributed Power Generation System Using a Tunable-Q Wavelet Transform and an Artificial Neural Network. *Energies* **2020**, *13*, 4238. [CrossRef]
33. Frizzo Stefenon, S.; Zanetti Freire, R.; dos Santos Coelho, L.; Meyer, L.H.; Bartnik Grebogi, R.; Gouvêa Buratto, W.; Nied, A. Electrical Insulator Fault Forecasting Based on a Wavelet Neuro-Fuzzy System. *Energies* **2020**, *13*, 484. [CrossRef]
34. Zhu, H.; Li, X.; Sun, Q.; Nie, L.; Yao, J.; Zhao, G. A Power Prediction Method for Photovoltaic Power Plant Based on Wavelet Decomposition and Artificial Neural Networks. *Energies* **2016**, *9*, 11. [CrossRef]
35. De Giorgi, M.G.; Campilongo, S.; Ficarella, A.; Congedo, P.M. Comparison Between Wind Power Prediction Models Based on Wavelet Decomposition with Least-Squares Support Vector Machine (LS-SVM) and Artificial Neural Network (ANN). *Energies* **2014**, *7*, 5251–5272. [CrossRef]
36. Kamal, T.; Karabacak, M.; Blaabjerg, F.; Hassan, S.Z.; Fernández-Ramírez, L.M. A Novel Lyapunov Stable Higher Order B-spline Online Adaptive Control Paradigm of Photovoltaic Systems. *Sol. Energy* **2019**, *194*, 530–540. [CrossRef]
37. Tan, K.T.; So, P.L.; Chu, Y.C.; Chen, M.Z.Q. Coordinated Control and Energy Management of Distributed Generation Inverters in a Microgrid. *IEEE Trans. Power Deliv.* **2013**, *28*, 704–713. [CrossRef]
38. Zhu, D.; Yang, R.; Hug-Glanzmann, G. Managing microgrids with intermittent resources: A two-layer multi-step optimal control approach. In Proceedings of the North American Power Symposium 2010, Arlington, TX, USA, 26–28 September 2010.
39. Chen, C.; Duan, S.; Cai, T.; Liu, B.; Hu, G. Optimal allocation and economic analysis of energy storage system in microgrids. *IEEE Trans. Power Electron.* **2011**, *26*, 2762–2773. [CrossRef]

Article

Development of Home Energy Management Scheme for a Smart Grid Community

Md Mamun Ur Rashid [1,2], Fabrizio Granelli [1], Md. Alamgir Hossain [3,4,*], Md. Shafiul Alam [5], Fahad Saleh Al-Ismail [5], Ashish Kumar Karmaker [4] and Md. Mijanur Rahaman [4]

1 Department of Information Engineering and Computer Science, University of Trento, 38122 Trento, Italy; mdmamunur.rashid@alumni.unitn.it or mamun@niter.edu.bd (M.M.U.R.); granelli@disi.unitn.it (F.G.)
2 Department of Electrical & Electronic Engineering, National Institute of Textile Engineering and Research (NITER), Dhaka 1350, Bangladesh
3 Capability Systems Centre, School of Engineering & Information Technology, University of New South Wales, Canberra 2612, Australia
4 Department of Electrical and Electronic Engineering, Dhaka University of Engineering & Technology (DUET), Gazipur 1700, Bangladesh; ashish@duet.ac.bd (A.K.K.); mijan.duet@yahoo.com (M.M.R.)
5 K. A. CARE Energy Research & Innovation Center, King Fahd University of Petroleum & Minerals, Dhahran 31261, Saudi Arabia; mdshafiul.alam@kfupm.edu.sa (M.S.A.); fsalismail@kfupm.edu.sa (F.S.A.-I.)
* Correspondence: Md.Hossain6@unsw.edu.au

Received: 17 July 2020; Accepted: 13 August 2020; Published: 19 August 2020

Abstract: The steady increase in energy demand for residential consumers requires an efficient energy management scheme. Utility organizations encourage household applicants to engage in residential energy management (REM) system. The utility's primary goal is to reduce system peak load demand while consumer intends to reduce electricity bills. The benefits of REM can be enhanced with renewable energy sources (RESs), backup battery storage system (BBSS), and optimal power-sharing strategies. This paper aims to reduce energy usages and monetary cost for smart grid communities with an efficient home energy management scheme (HEMS). Normally, the residential consumer deals with numerous smart home appliances that have various operating time priorities depending on consumer preferences. In this paper, a cost-efficient power-sharing technique is developed which works based on priorities of appliances' operating time. The home appliances are sorted on priority basis and the BBSS are charged and discharged based on the energy availability within the smart grid communities and real time energy pricing. The benefits of optimal power-sharing techniques with the RESs and BBSS are analyzed by taking three different scenarios which are simulated by C++ software package. Extensive case studies are carried out to validate the effectiveness of the proposed energy management scheme. It is demonstrated that the proposed method can save energy and reduce electricity cost up to 35% and 45% compared to the existing methods.

Keywords: home energy management scheme (HEMS); smart grid; renewable energy sources (RESs); power sharing algorithm (PSA); residence energy management (REM)

1. Introduction

Energy, a scare resource that needs to be efficiently utilized, is considered the most indispensable component in modern society. For a typical home, energy losses are generally higher due to line losses increased by a factor of double [1]. These energy losses can be minimized by developing an efficient home energy management scheme (HEMS) that plays a crucial role in our daily life due to increasing energy demand [2]. The minimization of energy cost and the maximization of user comfort are the typical optimization problems in recent smart homes [3,4]. The authors in research [5] illustrate an aggregator based household energy scheduling plan in the smart grid framework where all home appliances are managed by different power scheduling schemes. However, the backup battery storage system (BBSS) is not considered in that research. Electricity consumers are not permitted to access the grid side mechanisms due to the high voltage grid and sophisticated device operation. They can manipulate on energy consumption to minimize the peak demand while reducing electricity bills [6,7]. To decrease electricity costs of residential users (RUs), it is important to maintain an energy scheduling plan for electrical appliances concerning the price of the respective time slot. The energy management of home appliances deals with utility tariffs and usages patterns [8]. Conversely, the smart home energy management scheme within the smart grid environment can integrate the alternative sources of energy and optimize the monetary cost [9]. Authors in research [10] illustrate the impact of the change in users' behavior in the smart grids framework. The smart consumption and production behaviors are analyzed in this study to assist smart grid diffusion. The energy management is highly dependent on demand response in a smart grid framework which is illustrated by evolutionary analysis in [11].

Residential utility management systems have been developed and validated through simulations and several case studies [12,13]. However, they did not consider the satisfied power demand and power-sharing algorithm to effectively handle the residential consumer's power demand. Researchers have taken some initiatives to manage a range of household appliances based on energy demand pattern. Most of the literature is related to minimizing the energy cost with carbon emissions and the impact of renewable energy integration in the smart grid [14]. A real-time energy management problem was solved using a modified particle swarm optimization to investigate the nominal battery state of a microgrid environment while minimizing operating charge. This work was extended to consider the degradation cost of the battery for a two-step ahead energy management scheme [4]. In [4], the authors demonstrated battery degradation costs which solely depends on a day-ahead energy scheduling and on a charging-discharging patterns. The optimization problem is solved using a framework consisting of Rainflow algorithm, particle swarm optimization (PSO), and scenario techniques. However, in these papers, the power-sharing algorithm and backup battery storage system are not taken into account. The authors of [15,16] provided a compact assessment of home energy management systems (HEMS). They explained the initial challenges of HEMS and then outlined some insights on the existing literature regarding the modeling of demand-side management (DSM) followed by wireless communication infrastructure. An optimization technique is developed through several programming strategies to handle home appliances with a solar system installed at the premises [17]. However, the impacts of different load patterns such as peak, off-peak and, mid-peak loads on renewable energy and BBSS are not considered. A fuzzy logic-based home energy management system (HEMS) is developed by the authors in [18] which is more effective compared to [19,20] but the lowering of energy cost by applying power-sharing algorithm is not taken into account. The home batteries, monitoring technologies, and smart heat pumps are analyzed in a smart grid architecture to show how the householders participate in HEM [21].

While several works deal with the advanced methods, these methods are not typically adopted by industry due to the reliability and security issues of a power system. Therefore, in order to minimize energy consumption and cost for residential consumers, this research develops an efficient energy management scheme using commercial software, C++, considering RESs and a BBSS integrated for smart grid model. The proposed HEMS offers higher customer satisfaction with low energy costs compared to the existing method.

The contributions of this paper can be listed as follows:

- A computationally efficient simulation model for HEM is developed using the C++ software package.
- The proposed model enables a power-sharing strategy in a community being assessed sequentially.
- The power-sharing technique with the proposed model saves energy cost up to 35% and 45% compared to conventional techniques.
- Mathematical modelling is developed to facilitate extensive analysis.

The rest of the paper is organized as follows: Section 2 discusses the problem description, existing power management system, and smart grid architecture including advanced smart metering infrastructure. Section 3 explains the system model, including a solar generator, wind generator, energy storage systems, loads and utility grid. Mathematical modeling of a smart home and the proposed algorithms are also presented in the section. Analysis of the results and graphical representation of simulation results are elaborated in Section 4. Finally, the conclusion this research is depicted in Section 5.

2. Problem Description

In this section, the proposed smart grid architecture and its advanced features with the application of intelligent smart metering infrastructure in the smart grid framework are presented in this section.

Smart Grid Architecture

The rapid growth in household appliances has significantly increased the power demand of residential consumers. Nowadays, most electricity demand is met by fossil fuel-based generation system which heavily pollutes the environment by emitting greenhouse gas [22,23]. According to the international standards, each country must put a cap on the carbon emission to save the environment. Thus, researchers are working hard to figure out the new means of electricity generation. In this research, the renewable energy sources (RESs) integration is found to be one of the most costs effective and environment-friendly solutions [22,24]. Normally, residential loads consist of washing machine, freezer, fans, lights, laptops, television, heater, dishwasher, microwave woven, and so on. Most of the residential consumers suffer from load shading, unstable voltage levels, service line loss, higher utility tariff, unscheduled system maintenance, and so on. To overcome these limitations, it is important to build an effective energy optimization method. To address these issues, a smart grid architecture is proposed with an efficient home energy management scheme (HEMS) to provide a profitable and sustainable solution.

The proposed community-based smart grid architecture illustrated in Figure 1, consists of three different scenarios considering local utility supply, renewable energy sources (RESs) such as solar cell and windmill and backup battery storage system (BBSS). It has three different energy communities such as community A, B, and C. To identify residential consumer effectively these communities are considered. Moreover, we have considered three scenarios where scenario 1 considers a residence with local utility supply as the primary source of energy whereas scenario 2 considers renewable energy as the primary source of energy and local utility supply as the secondary source. Besides, scenario 3 uses renewable energy and backup battery storage system (BBSS) as a primary source and local utility as secondary source in case of scarcity. When any particular consumer request power and met by the energy community is called satisfied power demand. In a day, during any particular period if any user requests power to the community and receives sufficient power called satisfied power demand. An efficient power-sharing algorithm is developed to meet user power demand using battery power sharing among the community. The novelty of this research is power-sharing technique for the residential consumers within the community [25].

A backup battery storage system (BBSS) bank is used to tackle emergency cases while all other power supplies are off. The BBSS charging and discharing are performed based on real-time pricing in order to reduce daily operation cost. Each residence has local utility supply, solar and wind power generation, smart battery, and smart meter. The smart meters are connected with all smart home appliances via a wireless connection. The use of smart meters (SM) in smart grid facilitates to record the exchange of information between utility companies, and electricity consumption. Home appliance usages are managed by scheduling and switching on/off some of them and track consumption over a period [26].

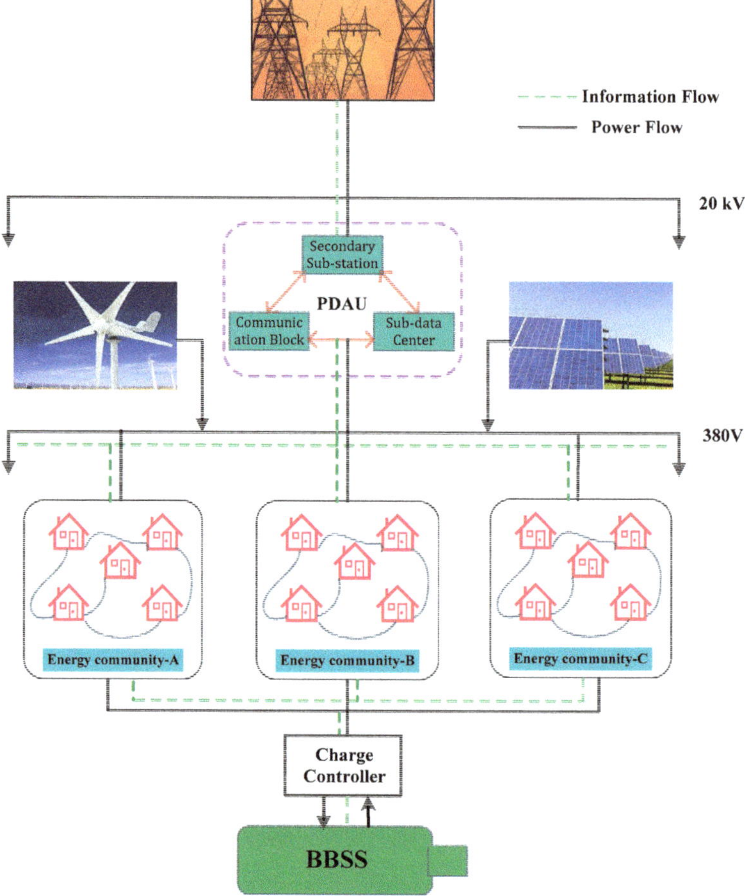

Figure 1. Proposed community-based smart grid architecture.

A smart home provides an excellent interface between household applicants and home appliances to keep energy use record and take essential and effective decisions [27]. All home appliances are connected via smart meters and permit a consumer to move energy use from peak-hours to mid-peak or off-peak hours [28]. It is capable of detecting the energy use charges in real-time by taking some electrical parameters.

The advanced smart metering architecture in the smart grid framework outlined in Figure 2 consists of metering and communication infrastructures. The SM unit includes time-of-use price monitoring, demand-side management, and automatic meter reading.

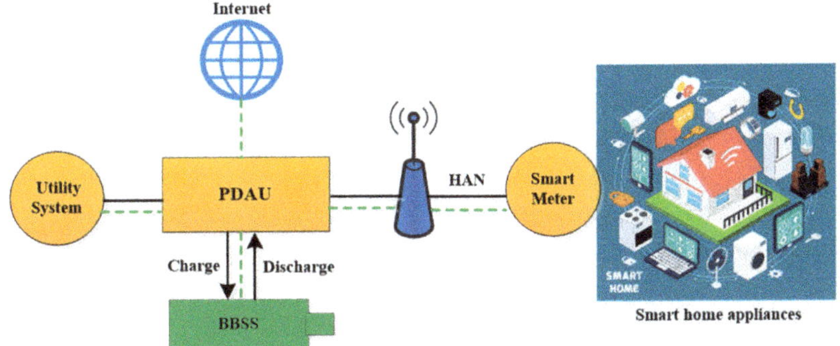

Figure 2. Advanced smart metering architecture in the smart grid framework.

The wireless communication technologies permit dual flow to enable the smart meter. Thus, smart meter permits exchange information with distant centers and to execute the control instructions. Multiple home appliances are connection with the smart meter in the residential premises by the aid of home area network (HAN). Wireless technologies such as ZigBee, Wi-Fi, and WiMAX [29,30] are effectively used in the smart grid environment. The benefits of advanced smart metering infrastructure are outlined below:

- Time-of-use pricing information can acquire from the distant consumer price indications.
- Customer's energy use information can accumulate, store and notify any particular time intervals or real-time.
- A detailed load patterns can develop the energy management process effectively.
- Smart meter can locate and identify the outages of any particular consumer by sharing a control message throughout the entire energy community.
- Power circuit can open or close over a long distance.
- Feasible to identify line losses and stealing exposure.

These meters send and receive information from the power data aggregator unit (PDAU) and permit residential consumers to handle their household appliances distantly. SMs of those communities are designed in such a way that allows them to be monitored by PDAU. If any users have extra demand, the smart meter sends a demand message to PDAU. After receiving the message, PDAU executes a power-sharing algorithm (PSA) to sequentially fulfill the user demand. Frist, PDAU forwards a demand message in the whole community. If the neighbor user can meet the extra demand, PDAU channels the excess stored energy of neighbors to the desired user. If demand cannot be fulfilled by neighbor users within the community, power will be acquired from the local utility companies. Moreover, if all power suppliers are unable to provide power then if any other user who wishes to receive power can take from BBSS by the permission of PDAU. PDAU is the main controller of this infrastructure. The BBSS keeps charging through the charge controller via a local utility supply while energy charges at low price intervals. Besides, BBSS can discharge when the community consumer energy demand is increased or all other supplies are off. To implement this framework, a simulator is designed and developed to evaluate the home user power consumption and cost considering satisfied power demand (hourly & daily basis) for one week. This method can save a significant amount of energy and cost compared to the existing systems.

3. System Modeling

This section deals with system components such as solar and wind generator, energy storage system, loads and utility grid, and a typical energy consumption profile for residential consumer's and home appliances considering peak-load, mid-peak, and off-peak loads and real-time electricity prices.

3.1. System Components

3.1.1. Solar Generator

Photovoltaic (PV) panels use sunlight to convert solar energy into usable DC electricity. The amount of PV power generated depends on the array size, solar radiance and solar insolation. Solar radiance is emphatically relying on surrounding climate and varies significantly as depicted in Figure 3. In order to gain maximum yield, PV panels are equipped with maximum power point tracking (MPPT) technology [31]. The performance of a PV system is assessed based on its output power. Output power depends on the size of the array and the overall efficiency of the installed PV system. PV power output as a function of solar irradiance and being operated at MPPT mode is defined using the following equation:

$$P_s = \eta_s * A * SI(1 + \gamma(t_0 - 25)) \tag{1}$$

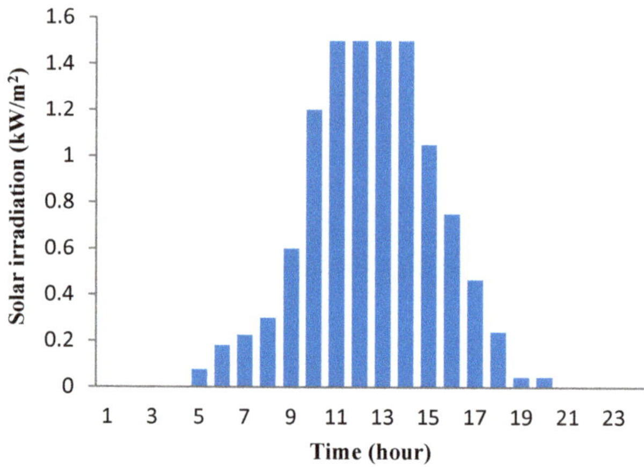

Figure 3. Predicted solar irradiation over a 24 h horizon.

In Equation (1) above, η_s represents efficiency, and A is the PV panel area. SI is the solar irradiation, t_0 is the air temperature, γ is the temperature coefficient of the maximum output power and is usually denoted as a negative percentage per °C or K. The PV technology and manufacturing parameters have great impact on γ, this study considers γ as −0.005/°C. The range of γ for silicon cells is between 0.004–0.006 per °C [32]. The expression of power for several solar PV panels is given below:

$$P_{sT} = P_s x N_s \tag{2}$$

where N_s is the number of solar generators.

3.1.2. Wind Generator

Wind power is a vital renewable energy resource that continues to thrive around the world due to features such as abundance, clean nature and being readily available. It is the method of power generation by rotating turbine blades installed at an elevated height. However, power generated from wind turbines is highly intermittent and stochastic in nature. It is highly dependent on the available wind speed and significantly varies based on the installation height. Figure 4 below illustrates wind speed over a range of time. The following power-law equation is used to convert wind speed recorded by anemometer (installed at tower heights) and transfer them based on hub heights [33]:

$$\frac{v}{v_0} = \left(\frac{h}{h_0}\right)^{\propto} \tag{3}$$

where velocity at hub height h_0 is v_0, velocity at hub height h is v, and the power-law exponent, α, depends on several parameters as presented in [34] which takes a value of 1/7 for open space. The wind turbine power as a function wind velocity can be represented as a piece-wise function below [35]:

$$P_w = \begin{cases} 0 & \text{if } v_f \leq v \text{ or } v \leq v_c \\ P_r \ * \frac{v^3 - v_c^3}{v_r^3 - v_c^3} & if\ v_c \leq v \leq v_r \\ P_r & if\ v_r \leq v \leq v_f \end{cases} \tag{4}$$

where the rated power is P_r, the wind rated speed is v_r, the cut-in wind speed is v_c, and the cut-off wind speed is v_f. Turbine manufacturers define a cut-in and cut-off speed for safety issues. The wind generator does not produce any power beyond the cut-off wind speed. The expression of total power for a number of wind turbines is given below [36]:

$$P_{wT} = P_w x N_w \tag{5}$$

where N_w is the number of wind generators.

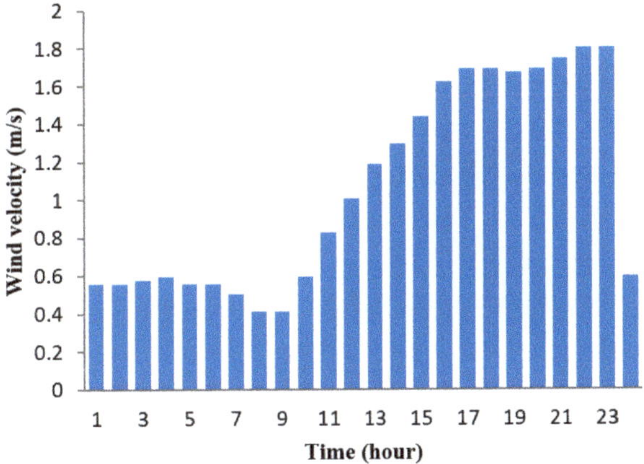

Figure 4. Predicted wind velocity over a 24 h horizon.

3.1.3. Backup Battery Storage Systems (BBSS)

The different types of energy storages, such as superconducting magnetic energy storage, batteries, flywheel energy storage, and compressed air energy storage, are studied in power system. Each of the energy storage devices has different features such as high-power density, high-energy density, high response time and so on. In this study, electrochemical batteries are used due to its high energy density. The charging and discharging of ESS can reduce the effect of the unpredictable nature of renewable energy sources (RESs) on the smart grid to balance power generation and demand. The expression for charging and discharging of battery is given as below"

$$BL(t) = BL(t-1) + \Delta t P_c(t) \eta_c \quad \text{for charging} \tag{6}$$

$$BL(t) = BL(t-1) + \Delta t P_d(t) \eta_d \quad \text{for discharging} \tag{7}$$

Subject to the following constraints:

$$P_{c,max} > P_c > 0$$

$$P_{d,max} < P_d < 0$$

$$BL_{max} > BL(t) > BL_{min}$$

where the battery charging power at time t is $P_c(t)$, the battery discharging power at time t and $P_d(t)$, the energy storage at time t is $BL(t)$, the time interval is Δt, the charging efficiency is η_c, and the discharging efficiency is η_d. This study considers η_c and η_d as unity for simplicity.

3.1.4. Loads and Utility Grid

The load profiles of residential users are illustrated in Figures 5–7. There are three different load patterns shown in separate figures. The maximum load is approximately 0.43 kW during peak-hours, the mid-peak hour is 0.42 kW during intermediate-hours and the minimum load is 0.32 kW during off-peak hours.

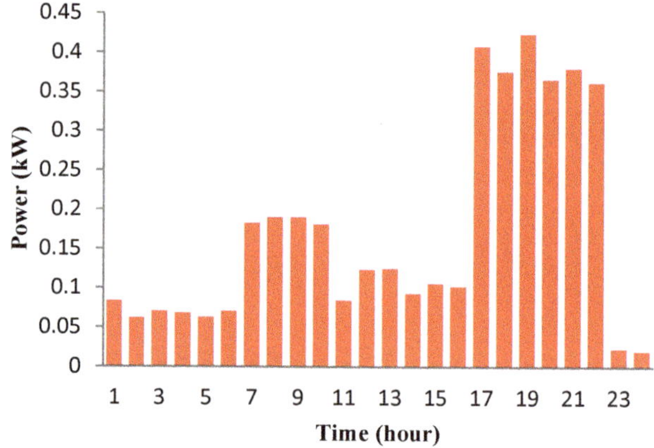

Figure 5. Predicted load profile of a house during peak hours over a 24 h horizon.

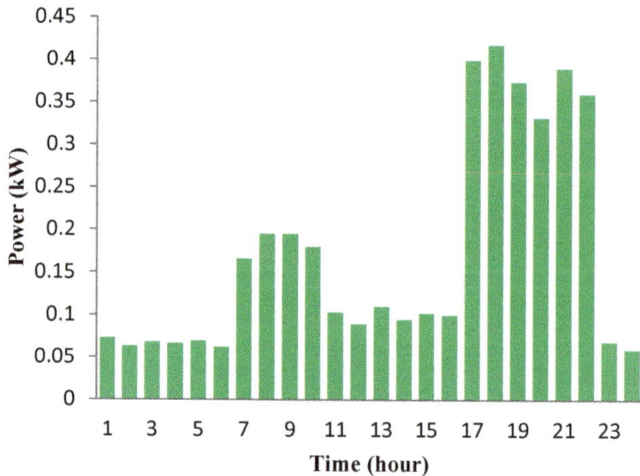

Figure 6. Predicted load profile of a house during mid-peak hours over a 24 h horizon.

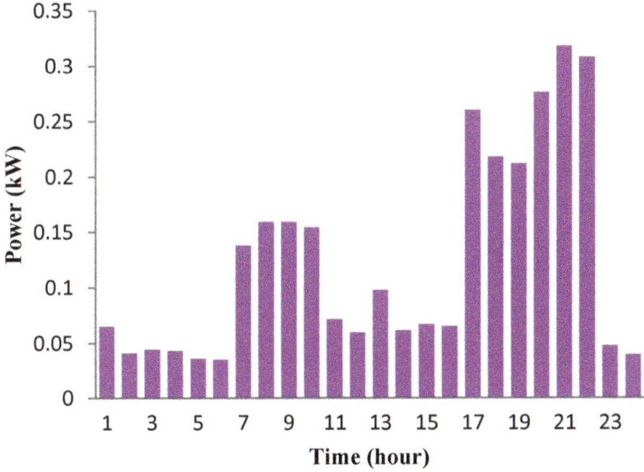

Figure 7. Predicted load profile of a house during off-peak hours over a 24 h horizon.

The estimated loads can meet the primary home appliances such as washing machine, freezer, fans, lights, laptops, television, heater, dishwasher, microwave woven etc. Home user energy demand patterns are fluctuating with time. It is assumed that the energy demand profiles are same for all residential consumers. Several measurements output values can shape the diverse energy demands for individual consumer.

Figure 8 indicates the real-time electricity prices for energy purchase per unit that is taken as a grid model in this study. These dynamic prices help electricity customers to schedule flexible loads, such as water heaters and washing machines. Electricity prices are generally announced an hour ahead from a distribution management company. We have considered a community smart grid that is connected to the grid.

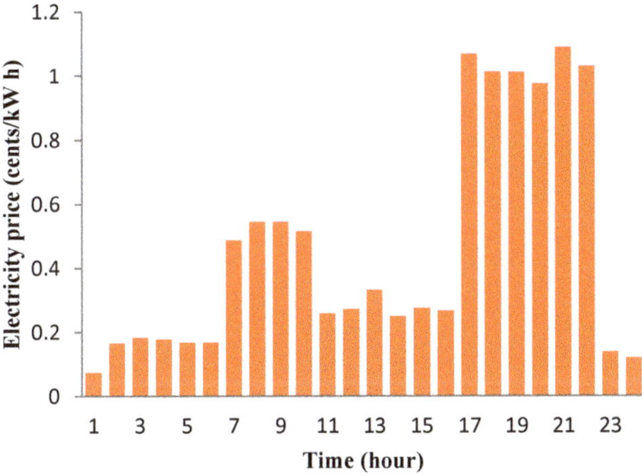

Figure 8. Predicted average electricity prices of a house over a 24 h horizon.

We have considered that the project is executed in an environment where building renewable energy plant is encouraged by the local government and the generated power can exchange to the utility grid without any transfer threshold. This process is due to insufficient power production, leading to low voltage outlines in the smart grid energy community. As a result, renewable energy owner can add more money on their utility business to help community inhabitants as well as to get profits.

3.1.5. Mathematical Modeling of a Smart Home for the Proposed Power-Sharing Algorithm

In this section, we analyze and model the electrical smart home appliances with renewable energy sources (RESs), backup battery storage system (BBSS), and related mathematical parameters in detail. The mathematical model contains different parts such as photo-voltaic system, small scale windmill, smart battery, backup battery storage system, utility tariffs, home appliances and their expenditure, etc. Renewable energy is generated by the solar cells and windmill that store in a smart battery. Total local renewable energy generation is formulated by the given equation:

$$P^{B} = p^{PV} + p^{W} \tag{8}$$

where P^{B} is the battery power storage, P^{PV} and P^{W} denote the solar photovoltaic and wind power, respectively. The power generation by photovoltaic solar cell, P^{PV}, and wind mill, P^{W}, are directly stored in a smart battery which is outlined by P^{B}. The total capacity of the battery is greater than sum of solar and wind power in order to save the battery from overcharge.

A single residential consumer all smart home appliances energy consumption is defined by the given equation:

$$E^{THA} = \sum (E^{L} + E^{WM} + E^{R} + E^{C} + E^{DW} + E^{H}) \tag{9}$$

where E^{THA} denotes the total energy of home appliances, E^{L} indicates lighting load, E^{WM} illustrates washing machine, E^{R} presents refrigerator load, E^{C} denotes computer load, and E^{DW} is the energy taken by dishwasher and E^{H} is the heating load.

All home users considered renewable energy as a primary source and utility as a secondary source.

$$E^{U} = (E^{B} + E^{BB}) - E^{THA} \tag{10}$$

where E^{U} represents the user energy consumption, E^{B} is the battery energy and E^{BB} is the backup battery energy. Similarly, E^{THA} indicates the total home appliance's energy of residential consumers. The user energy consumption is equal to the sum of battery and backup battery storage minus the total home appliances energy. Home users extra power demand is meet by the utility system and it is defined by the given Equation (11) and extra energy is formulated by the Equation (12).

$$E^{EE} = \sum (E^{UO} - E^{U}) \tag{11}$$

$$E^{EEC} = \sum ((E^{UO} - E^{U}) \times E^{UC})) \tag{12}$$

E^{EE} represents the extra energy, E^{UO} is the energy of utility operator and E^{U} is the user power. Similarly, E^{EEC} defines the extra energy cost and E^{UC} illustrates the utility cost. We have considered four user energy communities and the total energy consumption are calculated by the given equation:

$$E^{TUC} = \Sigma_{i=1}^{n} E^{TU} = (E^{U1} + E^{U2} + E^{U3} + E^{U4}) \tag{13}$$

where n equals to 4 (the total number of user energy communities) and E^{TU} represents the total user energy. We have considered a group of twenty home users in our proposed community smart grid and the total energy consumption in a community is given by the following equation:

$$E^{TUE} = \Sigma_{i=1}^{n} E^{UE} = \Sigma(E^{U1} + E^{U2} + E^{U3} + E^{U4} \ldots\ldots\ldots\ldots + E^{U\_N}) \tag{14}$$

where E^{TUE} represents the total user energy and E^{U} denote the individual user energy. Total users energy cost in a community is carried out by the given equation:

$$E^{TUEC} = \sum (E^{U1} + E^{U2} + E^{U3} + E^{U4} \ldots \ldots \ldots + E^{U\_N}) \times E^{UC} \tag{15}$$

where E^{TUEC} represents the total user energy cost and E^{U} denote the individual user energy and E^{UC} illustrates the individual user energy cost. The total number of hours in a day is given by the given equation which is 24 h.

$$H^{TNH} = \Sigma_{i=1}^{T} E^{H} \tag{16}$$

where T equals to 24 due to 24 h in a day and H^{TNH} represents the total number of hours and E^{H} denote the hourly user energy consumption. Total energy consumption and cost for a week is shown in the given equation:

$$D^{TND} = \Sigma_{i=1}^{W} E^{D} \tag{17}$$

where W equals to 7 and D^{TND} represents the total number of day and E^{D} denotes the daily user energy consumption. Total satisfied power demands are carried out by the given Equation (18)

$$P^{TSPD} = \sum P^{SPD} \tag{18}$$

where P^{TSPD} denotes the total satisfied power demand and P^{SPD} represents the satisfied power demand. However, the sum of individual satisfied power demand is equal to the total satisfied power demand.

3.2. The Proposed Algorithm and Simulations

In this section, we introduce the simulation framework and proposed a power-sharing algorithm. An advanced and efficient power-sharing algorithm is proposed which defines the step by step simulation execution process and priority basis energy management for residential consumers is placed in the Appendix A.

The Proposed Power-Sharing Algorithm (PSA)

The real-time electricity prices are considered in the simulation process, allowing us to overcome the conventional optimization problems. To resolve these issues an efficient algorithm is designed called power-sharing algorithm (PSA). The PSA is used to optimize the new proposed community-based smart grid model considering a backup battery storage system (BBSS) [37]. In this study PSA is also discussed to effectively share user battery power and backup energy storage within the community smart grid. The prime objective of the proposed algorithm is to minimize the energy cost with a high level of customer satisfaction. Residential consumers consume most of the energy to meet their smart home appliances. When smart battery and local utility are unable to support, a BBSS is connected with the power data aggregator unit (PDAU) to support the electricity supply. BBSS channels the energy to the respective user by the coordination of PDAU. All decisions including emergency case are made by PDAU. If the residential user (RU) energy consumption is greater than the battery capacity then battery needs to charge. At that time the user sends a demand message to the PDAU. Taking approval of PDAU, the user can receive shared energy from other user battery. Moreover, when all supplies are off, the requested user receives power from the BBSS service.

4. Simulation Results

The proposed energy management procedure and simulation results are outlined in this section to analyze and compare the effectiveness of the proposed HEM scheme. An on-grid tied smart grid framework is taken into account in this research. Sufficient PV solar and wind power generation reduce the utility tariffs. An efficient power-sharing algorithm (PSA) is developed to effectively manage residential consumer's smart battery power-sharing. We have also considered a backup battery storage system (BBSS) having a group of a battery bank to support residential consumers while all supply is off. The capacity of BBSS is selected as 30 kWh, and the minimum and maximum energy storage margins are taken as 5 kWh and 30 kWh, respectively.

The primary energy level of the BBSS is considered as 10 kWh. The smart grid constraints and the other input values are presented in Table 1. The input data are picked from real-time investigations. However, those are a bit scaled up or down due to precise application of this research. We have considered three smart grid scenarios, one existing and two proposed to compare the percentage of energy and cost savings. The simulation results are used to compare with the existing scenarios to show how the proposed scheme is better for reducing energy demand and cost significantly. It also evaluates the hourly and daily energy consumption and prices considering satisfied power demand to make the comparison effectively to propose an economic solution for residential consumers.

Table 1. Simulation parameters.

Parameters	Value	Unit
Solar PV system		
Total area, A	125	m^2
Efficiency, η_s	16	%
Maximum power	20	kW
Wind generators		
Cut in velocity	3	m/s
Cut out velocity	25	m/s
Rated speed	10	m/s
Maximum power	20	kW
Battery		
Initial energy level, BLo	10	kWh
Maximum energy level, BLmax	30	kWh
Minimum energy level, BLmin	5	kWh
Energy capacity	30	kWh
Total hour	24	h
Total day	7	d
Total user	20	
Total iteration	10	
Energy cost	0.072	€/kWh

4.1. Case Study

Several case studies are carried out to presents the residential consumer's energy consumption patterns and compare due to get economic solutions. A group of 20 residential consumers is considered to have several smart grid communities to evaluate energy and cost through some case studies. Besides, we have considered three scenarios one existing and two proposed. In scenario-1, all residential consumers have power supply from local utility only whereas in scenario-2, all residential consumers have a power supply both from the local utility and RESs. Moreover, in scenario-3, all residential consumers have power supply from local utility, renewable energy and, backup battery storage system (BBSS).

- Scenario 1: Residences with local utility support only;
- Scenario 2: Residences with local utility support and RESs;
- Scenario 3: Residences with local utility support, RESs and BBSS.

4.1.1. Case I

In this case, residential consumers considered a local utility as their primary source of energy. There is no renewable energy generator and energy storage system to support while power shortage. This approach is completely unable to manage economic issues and energy minimization due to the absence of RESs and backup battery storage systems (BBSS). All residence appliances receive energy from local utility and use as much as they require and after a certain time pay for the utility bills. Energy consumption and electricity cost management for three scenarios over 24-h time horizon is illustrated in Figure 9. In this research, we have considered scenario 1 as existing whereas scenario 2 is the proposed energy management scheme. Residential consumers have different load patterns such as peak-load, off-peak and mid-peak loads. The load profiles are presented in Figures 5–7. To optimize residential consumer's energy consumption and cost it is necessary to take some efficient initiatives and techniques.

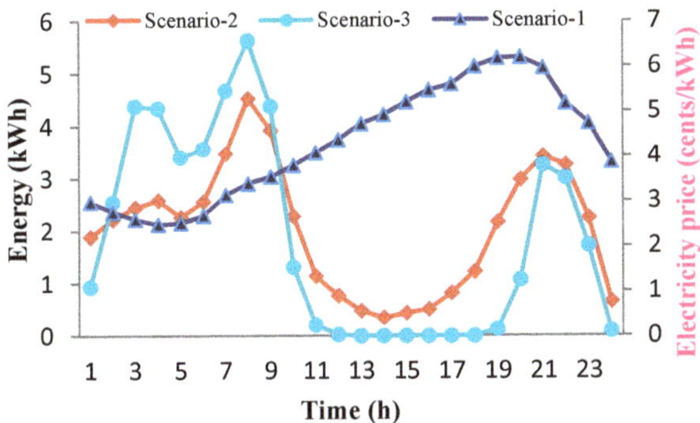

Figure 9. Energy consumption and electricity cost management for three scenarios over 24 h time horizon.

Some effective strategies and RESs integration are taken in to account in the proposed scheme. From Figure 9 it is clear that the highest energy consumption is attained by scenario 1 and the lowest energy consumption is cut by scenario 3.

4.1.2. Case II

A typical power production using solar panel and wind turbine is considered that are figured in Figures 3 and 4, respectively. The spasmodic power production from wind and solar generators is completely stochastic. Sometimes it is tough to meet residential consumer power demand due to unpredictable power generation. To overcome these issues a backup battery storage system (BBSS) is introduced in the smart grid environment. The comparison of proposed and existing energy and cost management over 24 h time horizon is presented in Figure 10 and the relationship of proposed and existing energy, and cost management for case II is tabulated in Table 2.

Table 2. Comparison of the proposed and existing energy, and cost management for case-II.

Scenarios	Energy (kWh)	Cost (EUR)	% Saving
Existing	77	5.50	0
Proposed 1	50	3.60	35

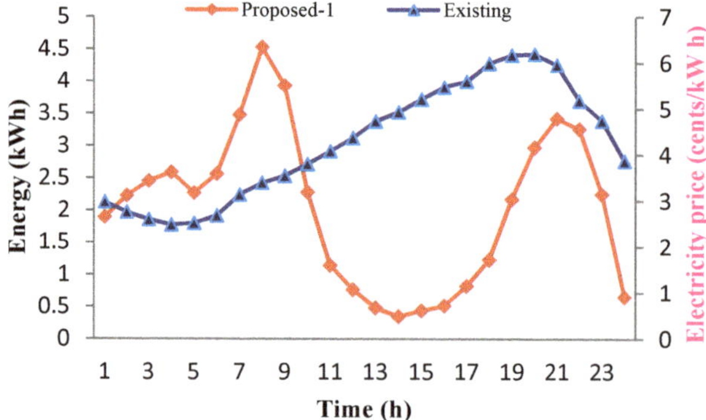

Figure 10. Comparison between proposed-1 and existing scenario of energy and electricity cost management over 24 h time horizon.

The comparison between proposed 1 and existing energy and cost management scheme over 24-h time horizon is presented in Figure 10 and solar and wind power generation is presented in Figures 3 and 4, respectively. Renewable energy generation is stored in a smart battery thus residence consumers can use and share battery power when they require. The electricity cost for this case study is 3.60 EUR, which is 35% less than the existing approach. It should be mention that the cost-saving is high due to the power purchasing from grid and RESs power generator. Moreover, during the peak hours, RESs generates the extra power than the exact demand and it is an extra reason behind the dramatic increase in profits.

Satisfied power demand: When residential consumers can not satisfy power demand by local utility supply then RESs can support power demand using an efficient power-sharing algorithm (PSA). Renewable energy generation is stored in a smart battery and residential consumers can use and share their battery power each other when they require. Figure 11 presents the satisfied power demand management between existing and proposed scenarios over 24 h time horizon.

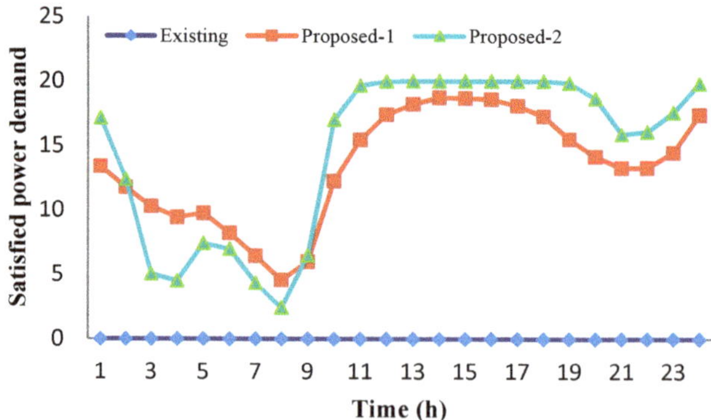

Figure 11. Satisfied power demand management between existing and proposed scenarios over 24 h time horizon.

Figure 12 presents the satisfied power demand management between existing and proposed 1 scenario over 24 h time horizon. In the case of scenario 1, the total number of satisfied power demand is empty due to local utility supply only. However, scenario 2 has both local utility and renewable energy generation support, for this reason, it can satisfy 18% power demands and tabulated in Table 3.

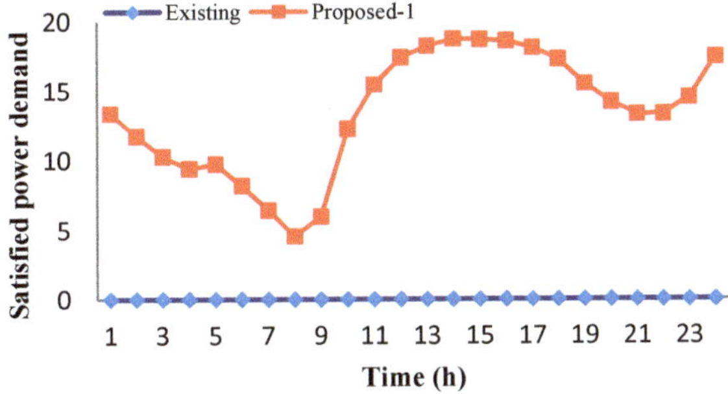

Figure 12. Comparison of satisfied power demand management between proposed 1 and existing scenario over 24 h time horizon.

Table 3. Comparison of the proposed and existing satisfied power demand management for case-II.

Scenarios	Satisfied Power Demand	% Saving
Existing	0	0
Proposed 1	16	18

4.1.3. Case III

To validating the efficiency of the offered method, the simulation procedure is executed to evaluate the daily energy and cost considering a backup battery storage system (BBSS) and power generation from RESs. The day with RESs power generation and BBSS is added to the simulation to notice the retort of the energy and cost minimization. Figure 13 presents the energy consumption and electricity cost management for three scenarios over 24 h time horizon. The electricity charge of the surviving method is 3.60 EUR with a day simulation period. 24-h simulation is observed in the proposed energy management approach where the BBSS is charged from the utility grid up to 30 kWh. Besides, during the high price hours of the electricity, BBSS fully discharged to fulfill the power demand. Again the BBSS consumed electricity for charging itself from the grid during the low price hours and touches the highest level of energy. A comparison between the offered approach and the existing ones is revealed in Table 4. The BBSS takes charge from the RESs and reaches the highest levels of energy in the subsequent time slots. The proposed 2 scenario can save energy and cost up to 45% compared to the existing method.

Satisfied power demand: When residential consumers can not satisfy power demand by local utility supply then an efficient power-sharing algorithm starts to satisfy user power demand using renewable energy sources and BBSS. Figure 14 presents the comparison of the satisfied power demand management between existing and proposed 2 scenarios over 24 h time horizon. In the case of the existing scenario, the total number of satisfied power demand is empty due to local utility supply only. However, the proposed 2 scenario has local utility, renewable energy generation, and BBSS support, for this reason, it can satisfy the highest numbers of power demands which are 22.5% and demonstrated in Table 5.

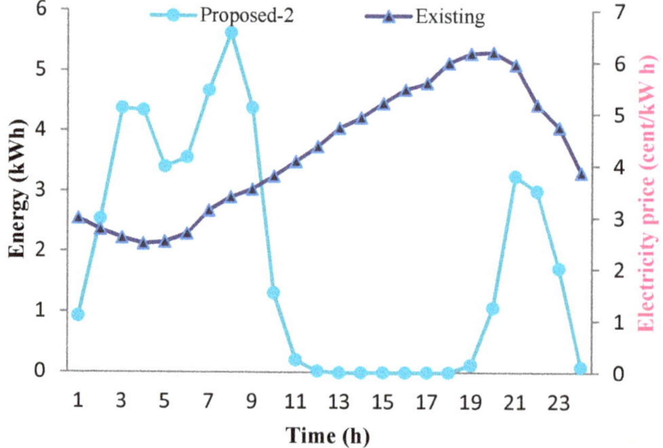

Figure 13. Comparison between proposed-2 and existing scenario of energy and electricity cost management over 24 h time horizon.

Table 4. Comparison of the proposed and existing energy, and cost management for case-III.

Scenarios	Energy (kWh)	Cost (EUR)	% Saving
Existing	50	3.60	0
Proposed 2	42	3.00	45

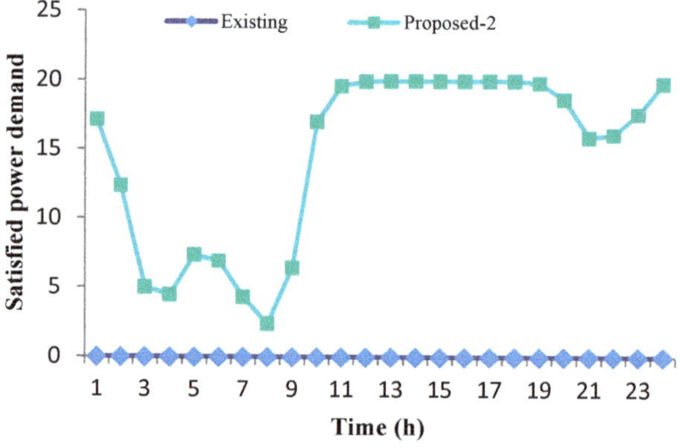

Figure 14. Comparison of satisfied power demand management between proposed-2 and existing scenario over 24 h time horizon.

Table 5. Comparison of the proposed and existing satisfied power demand management for case-III.

Scenarios	Satisfied Power Demand	% Saving
Existing	0	0
Proposed 2	20	22.5

4.2. Discussion

In this section, the overall energy and cost minimization for the residential consumers is outlined based on several case studies. Case I is completely unable to minimize energy and monetary cost due to the absence of an alternative sources of energy and BBSS. However, case II and case III can save a significant amount of energy and cost considering renewable energy, BBSS and satisfied power demand. Percentage of savings is tabulated in Table 6 to identify the most profitable case.

Table 6. Comparison of the proposed and existing energy, and cost management within case-I, II, and III.

Cases	Energy (kWh)	Cost (EUR)	% Saving
Case I	77	5.50	0
Case II	50	3.60	35
Case III	42	3.00	45

Figure 15 presents the percentage of energy savings management within different case studies over 24 h time horizon. Case II can save 35% energy consumption whereas case III can save 45%. The pink line indicates the percentage of energy savings which is sharply increasing due to cost-effective energy management.

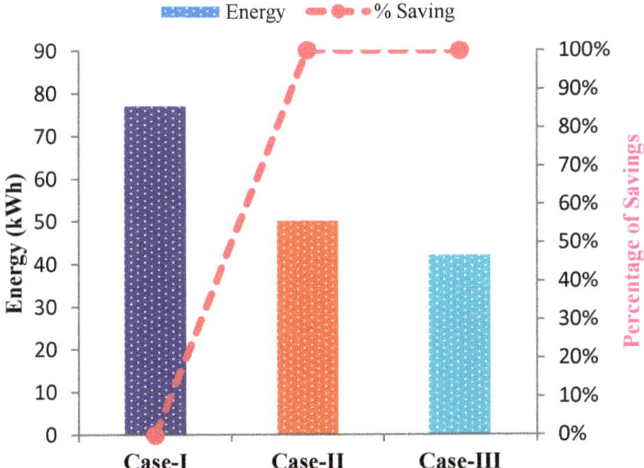

Figure 15. Percentage of energy savings management within case-I, II, and III over 24 h time horizon.

On the other hand, Figure 16 illustrates the percentage of monetary cost savings management within different case studies over 24 h time horizon. Case II can save 35% energy cost whereas case III can save 45%. The pink line indicates the percentage of energy cost savings which is linearly increasing due to economic energy optimization.

The comparison of average daily energy consumption and cost management within three different cases are tabulated in Table 7. From Figure 17 the daily energy consumption is the greatest in case I whereas in case II and case III is declining. Evaluating several case studies the overall energy and cost reduction is profitably accelerated in case II and case III due to considering RESs, BBSS, and PSA in the smart grid environment. To get dual supply opportunity we have considered an on-grid tied smart grid system.

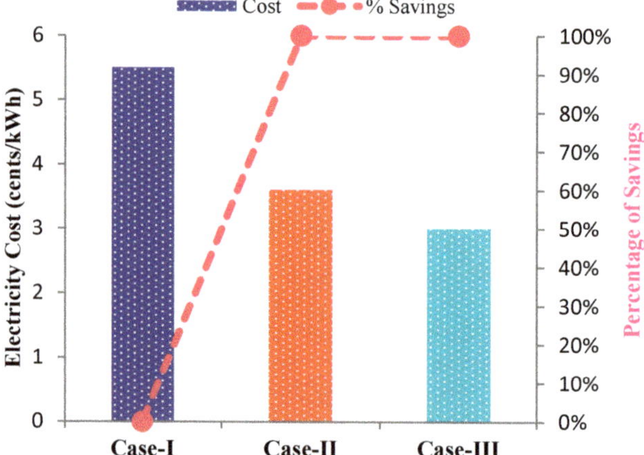

Figure 16. Percentage of energy cost savings management within case-I, II, and III over 24 h time horizon.

Table 7. Comparison of the average daily energy consumption and cost management within three cases.

Cases	Energy (kWh)	Cost (EUR)
Case I	77	5.50
Case II	50	3.60
Case III	42	3.00

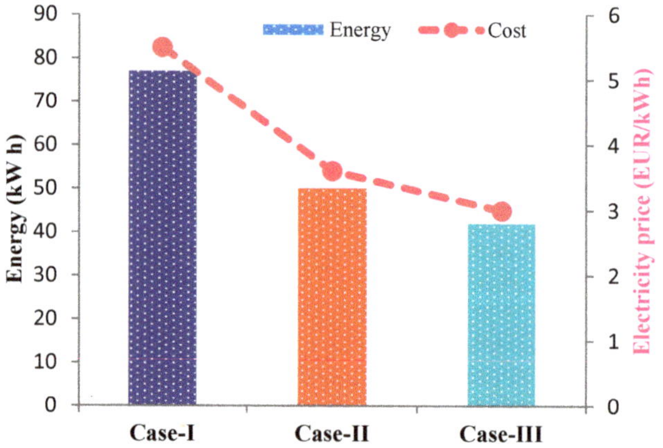

Figure 17. Comparison of average daily energy consumption and cost management within case-I, II, and III.

In case any load shading due to system maintenance then RESs and BBSS can support a part of the energy community. Sometimes residential consumers cannot estimate load management properly due to unpredictable RESs generation. However, the outlined efficiency is not much due to RESs and BBSS limitations. A large BBSS and the best quality of RESs can significantly increase the percentage of energy and cost savings. However, the residential consumers may not be interested or effort the cost due to huge primary investment.

A comparison of energy consumption and cost for a single residential consumer in a day is illustrates in Figure 18 and Table 8 to get a basic idea regarding energy and respective cost within different scenarios. The highest energy consumption shows in scenario-1 whereas the lowest energy consumption is done by the scenario 3.

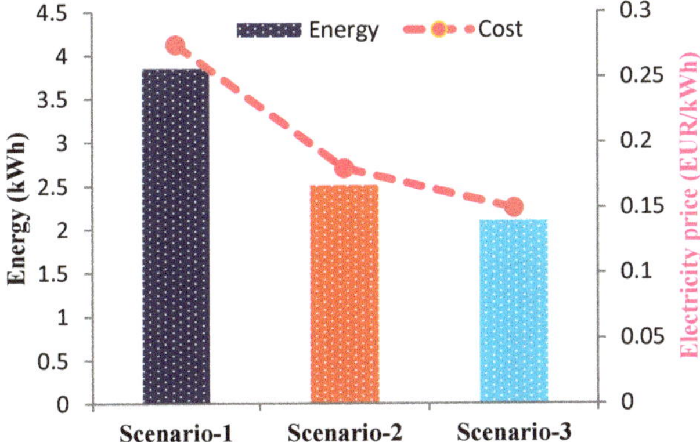

Figure 18. Comparison of the average daily energy consumption and cost management for a single user within three scenarios.

Table 8. Comparison of the average daily energy consumption and cost management for single user within three scenarios.

Scenarios	Energy (kWh)	Cost (EUR)
Scenario 1	3.85	0.27
Scenario 2	2.50	0.18
Scenario 3	2.10	0.15

A comparison of the proposed and existing satisfied power demands within different case studies are illustrated in Figure 19 and tabulated in Table 9. In Figure 19 the pink line indicates the percentages of satisfied power demand savings which are 18% and 22.5%, respectively. Case III has 4.5% more satisfied power demand savings than case II and 22.5% more savings than case I. The prescribed percentage of savings is not much and the amount can be accelerated with the aid of a large backup battery bank and RESs system. However, it requires large space and huge primary investment. For this reason, the residential consumers may not effort or interested to invest a large amount of economy. Moreover, the proposed model can be implemented effectively on a small scale micro-grid environment.

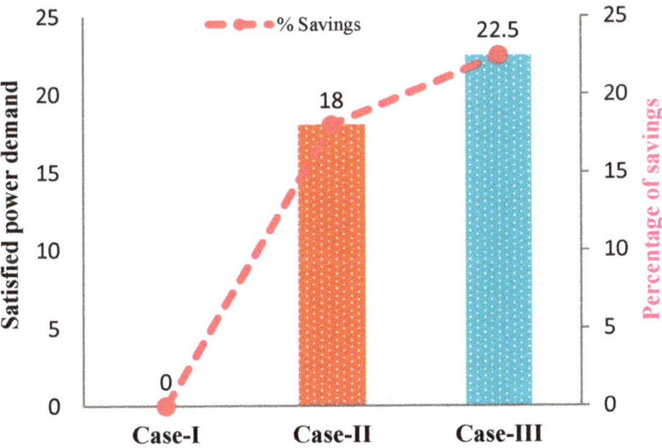

Figure 19. Percentage of satisfied power demand savings within cases I, II, and III over 24 h time horizon.

Table 9. Comparison of the proposed and existing satisfied power demands within case-I, II and III.

Cases	Satisfied Power Demand	% Saving
Case I	0	0
Case II	16	18
Case III	20	22.5

4.3. Case Comparison

The vital concern of this research is the case comparison to understand and outline the most cost-effective solution to minimize energy and cost for the residential consumers. Case I consider fixed utility tariffs and a nominated kW demand thus the consumer cannot exceed the allocated energy demand which is set by the utility companies. For this reason, the consumer cannot avoid the peak load and higher utility tariffs. Case II demonstrated an idea to overcome these difficulties by installing RESs with a utility grid to compensate the peak load and higher utility costs. However, it is not always possible due to dependency on solar irradiation and wind velocity during cloudy weather and winter season. The compared amount is not much higher due to the moderate size of ESS. However, installing a large BBSS requires a big battery bank, leading greater energy costs which consumers may not effort.

Finally, case III has additional backup battery storage systems (BBSS) compared to case II to eliminate the mentioned difficulties which is presented in Figure 20. The BBSS is charged during low energy cost periods and discharges when peak loads and extra demand arise. We have taken into account the point of view that the initial investment for RESs and BBSS is cost-effective and sustainable. A depreciation cost is considered with RESs and BBSS to make the proposed model more realistic. Government and many energy providers encourage residential consumers by providing home loans for installing small scale renewable energy generators. This may be a flexible opportunity for low-income bracket individuals and mid-category consumers. The proposed model can be implemented effectively on a small scale micro-grid environment.

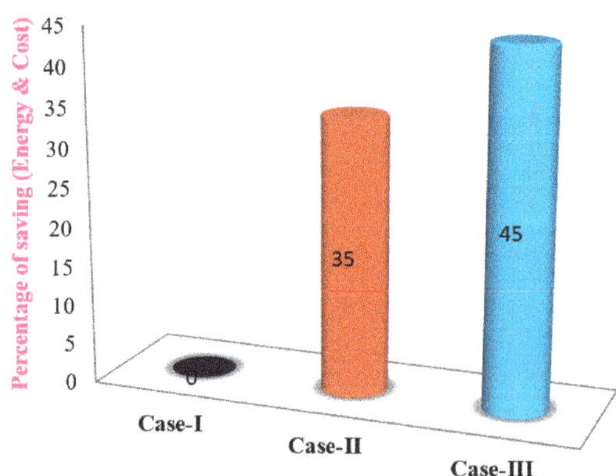

Figure 20. Comparison of the percentage of energy and cost saving for cases I, II and III.

5. Conclusions

A home energy management scheme for smart grid community with renewable energy sources and backup battery storage system is developed in this paper for a group of residential consumers. The residential energy management (REM) problem with consumer's energy consumption, cost minimization, and satisfied power demand is modeled. The home appliances' priorities, energy availability and real-time energy pricing are considered in the proposed power-sharing technique. The offered model with the power-sharing technique is implemented in a practical system considering twenty residential consumers. The economic benefits of the proposed REM scheme with BESS are investigated with three different scenarios. The financial analysis of the proposed technique is conducted to show the monetary savings which is guaranteed in our proposed scheme. It is clear from the simulation results that the recommended scheme can save energy and costs by up to 35% and 45% compared to existing methods. Also, the results outline that the proposed method can save satisfied power demands by up to 16% and 22.5% compared to the existing method. From overall analysis, the proposed REM model can enhance the benefits of the RESs and BBSS integration. Moreover, the simulation outcomes prove the potential benefits of the residential consumers and endorse the green effort to improve the sustainability and effectiveness of the power management infrastructure.

Author Contributions: M.M.U.R., F.G., M.A.H. and M.S.A. initiated the idea, designed the energy management scheme and drafted the article. M.M.U.R., F.S.A.-I., A.K.K. and M.M.R. performed the simulations and formal analysis. All authors made revisions of the articles. All authors have read and agreed to the published version of the manuscript.

Funding: This research received no external funding.

Conflicts of Interest: The authors declare no conflict of interest.

Nomenclature

SG	smart grid
SM	smart meter
HEMS	home energy management scheme
PDAU	power data aggregator unit
DSM	demand-side management
PSA	power-sharing algorithm
HANs	home area networks
SMA	smart metering architecture
DSO	distribution system operator
RESs	renewable energy sources
EMS	energy management system
ESS	energy storage system
SI	solar irradiation
PV	photovoltaic
P_wT	total wind power
P_sT	total solar power
BL_o	initial battery energy level
BL_{max}	maximum battery energy level
BL_{min}	minimum battery energy level
E^B	battery storage
p^S	solar power
REM	residence energy management
p^W	Wind power
E^{THA}	total home appliances
E^L	lighting load
E^{WM}	power by washing machine
W^R	power taken by refrigerator
E^C	power taken by computer
E^{DW}	power taken by dishwasher
E^H	power taken by heater
E^{BB}	backup battery storage
E^{EE}	users extra energy
E^{EEC}	users extra energy cost
P^{UC}	utility company
P^{UP}	utility price
C^T	total number of community
RU^T	total number of residential users
E^{TUE}	total users energy
E^{TUEC}	total users energy cost
H^{TNH}	total number of hours
D^{TND}	total number of days
p^{TSPD}	total satisfied power demand
I^T	total number of iteration
E^{CE}	cost of energy
B^{CSB}	capacity of smart battery

Appendix A

Algorithm A1: Power-Sharing Algorithm (PSA)

```
1:  Initialization.
2:  Load data.
3:  Start simulation.
4:  Iteration.
5:  Generate random values.
6:  if user power ≥ battery power then
7:          Charge user battery.
8:          Otherwise, the user battery is the maximum.
9:      end if
10: else if found healthy battery then
11:         Charge user battery.
12:         Otherwise, search for another healthy battery.
13:     end else if
14: if all power supply off then
15:         Use a backup battery storage
16:         Otherwise, use the user battery.
17:     end if
18:         Evaluate energy & prices.
19:         Evaluate satisfied power demand.
20: repeat step 4
21:         until iteration ≥ 10
22: Stop simulation
```

References

1. Farrokhifar, M.; Momayyezi, F.; Sadoogi, N.; Safari, A. Real-time based approach for intelligent building energy management using dynamic price policies. *Sustain. Cities Soc.* **2018**, *37*, 85–92. [CrossRef]
2. Sharifi, A.H.; Maghouli, P. Energy management of smart homes equipped with energy storage systems considering the PAR index based on real-time pricing. *Sustain. Cities Soc.* **2019**, *45*, 579–587. [CrossRef]
3. Sheikhi, A.; Rayati, M.; Ranjbar, A.M. Demand side management for a residential customer in multi-energy systems. *Sustain. Cities Soc.* **2016**, *22*, 63–77. [CrossRef]
4. Hossain, M.A.; Pota, H.R.; Squartini, S.; Zaman, F.; Guerrero, J.M. Energy scheduling of community microgrid with battery cost using particle swarm optimisation. *Appl. Energy* **2019**, *254*, 113723. [CrossRef]
5. Lezama, F.; Soares, J.; Canizes, B.; Vale, Z. Flexibility management model of home appliances to support DSO requests in smart grids. *Sustain. Cities Soc.* **2020**, *55*, 102048. [CrossRef]
6. Rahmani-Andebili, M. Scheduling deferrable appliances and energy resources of a smart home applying multi-time scale stochastic model predictive control. *Sustain. Cities Soc.* **2017**, *32*, 338–347. [CrossRef]
7. Yigit, K.; Acarkan, B. A new electrical energy management approach for ships using mixed energy sources to ensure sustainable port cities. *Sustain. Cities Soc.* **2018**, *40*, 126–135. [CrossRef]
8. Abushnaf, J.; Rassau, A.; Górnisiewicz, W. Impact of dynamic energy pricing schemes on a novel multi-user home energy management system. *Electr. Power Syst. Res.* **2015**. [CrossRef]
9. Reynolds, J.; Rezgui, Y.; Hippolyte, J.L. Upscaling energy control from building to districts: Current limitations and future perspectives. *Sustain. Cities Soc.* **2017**, *35*, 816–829. [CrossRef]
10. Perri, C.; Giglio, C.; Corvello, V. Smart users for smart technologies: Investigating the intention to adopt smart energy consumption behaviors. *Technol. Forecast. Soc. Chang.* **2020**, *155*, 119991. [CrossRef]
11. Liu, X.; Wang, Q.; Wang, W. Evolutionary Analysis for Residential Consumer Participating in Demand Response Considering Irrational Behavior. *Energies* **2019**, *12*, 3727. [CrossRef]

12. López, J.M.G.; Pouresmaeil, E.; Cañizares, C.A.; Bhattacharya, K.; Mosaddegh, A.; Solanki, B.V. Smart Residential Load Simulator for Energy Management in Smart Grids. *IEEE Trans. Ind. Electron.* **2019**, *66*, 1443–1452. [CrossRef]
13. Haider, H.T.; See, O.H.; Elmenreich, W. A review of residential demand response of smart grid. *Renew. Sustain. Energy Rev.* **2016**, *59*, 166–178. [CrossRef]
14. Karmaker, A.K.; Rahman, M.M.; Hossain, M.A.; Ahmed, M.R. Exploration and corrective measures of greenhouse gas emission from fossil fuel power stations for Bangladesh. *J. Clean. Prod.* **2020**, *244*, 118645. [CrossRef]
15. Beaudin, M.; Zareipour, H. Home energy management systems: A review of modelling and complexity. *Renew. Sustain. Energy Rev.* **2015**, *45*, 318–335. [CrossRef]
16. Hossain, M.A.; Pota, H.R.; Squartini, S.; Zaman, F.; Muttaqi, K.M. Energy management of community microgrids considering degradation cost of battery. *J. Energy Storage* **2019**, *22*, 257–269. [CrossRef]
17. Di Giorgio, A.; Pimpinella, L. An event driven Smart Home Controller enabling consumer economic saving and automated Demand Side Management. *Appl. Energy* **2012**, *96*, 92–103. [CrossRef]
18. Qayyum, F.A.; Naeem, M.; Khwaja, A.S.; Anpalagan, A.; Guan, L.; Venkatesh, B. Appliance Scheduling Optimization in Smart Home Networks. *IEEE Access* **2015**, *3*, 2176–2190. [CrossRef]
19. Erol-Kantarci, M.; Mouftah, H.T. Wireless sensor networks for cost-efficient residential energy management in the smart grid. *IEEE Trans. Smart Grid* **2011**, *2*, 314–325. [CrossRef]
20. Mahmood, A.; Khan, I.; Razzaq, S.; Najam, Z.; Khan, N.A.; Rehman, M.A.; Javaid, N. Home appliances coordination scheme for energy management (HACS4EM) using wireless sensor networks in smart grids. *Procedia Comput. Sci.* **2014**, *32*, 469–476. [CrossRef]
21. Smale, R.; Spaargaren, G.; van Vliet, B. Householders co-managing energy systems: Space for collaboration? *Build. Res. Inf.* **2019**, *47*, 585–597. [CrossRef]
22. Karmaker, A.K.; Ahmed, M.R.; Hossain, M.A.; Sikder, M.M. Feasibility assessment & design of hybrid renewable energy based electric vehicle charging station in Bangladesh. *Sustain. Cities Soc.* **2018**, *39*, 189–202. [CrossRef]
23. Alam, M.S.; Abido, M.A.Y.; El-Amin, I. Fault Current Limiters in Power Systems: A Comprehensive Review. *Energies* **2018**, *11*, 1025. [CrossRef]
24. Hossain, M.A.; Pota, H.R.; Squartini, S.; Abdou, A.F. Modified PSO algorithm for real-time energy management in grid-connected microgrids. *Renew. Energy* **2019**, *136*, 746–757. [CrossRef]
25. Ur Rashid, M.M.; Hasan, M.M. Simulation based energy and cost optimization for home users in a community smart grid. *Int. J. Renew. Energy Res.* **2018**, *8*, 1281–1287.
26. Lior, N. Sustainable energy development: The present (2009) situation and possible paths to the future. *Energy* **2010**, *35*, 3976–3994. [CrossRef]
27. McDaniel, P.; McLaughlin, S. Security and privacy challenges in the smart grid. *IEEE Secur. Priv.* **2009**, *7*, 75–77. [CrossRef]
28. Zipperer, A.; Aloise-Young, P.A.; Suryanarayanan, S.; Roche, R.; Earle, L.; Christensen, D.; Bauleo, P.; Zimmerle, D. Electric energy management in the smart home: Perspectives on enabling technologies and consumer behavior. *Proc. IEEE* **2013**, *101*, 2397–2408. [CrossRef]
29. Deng, R.; Yang, Z.; Chow, M.Y.; Chen, J. A survey on demand response in smart grids: Mathematical models and approaches. *IEEE Trans. Ind. Inform.* **2015**, *11*, 570–582. [CrossRef]
30. Yang, Z.; Chen, Y.X.; Li, Y.F.; Zio, E.; Kang, R. Smart electricity meter reliability prediction based on accelerated degradation testing and modeling. *Int. J. Electr. Power Energy Syst.* **2014**, *56*, 209–219. [CrossRef]
31. Xiao, W.; Lind, M.G.J.; Dunford, W.G.; Capel, A. Real-time identification of optimal operating points in photovoltaic power systems. *IEEE Trans. Ind. Electron.* **2006**, *53*, 1017–1026. [CrossRef]
32. Kaabeche, A.; Belhamel, M.; Ibtiouen, R. Sizing optimization of grid-independent hybrid photovoltaic/wind power generation system. *Energy* **2011**, *36*, 1214–1222. [CrossRef]
33. Justus, C.G. Wind energy statistics for large arrays of wind turbines (New England and Central U.S. Regions). *Sol. Energy* **1978**, *20*, 379–386. [CrossRef]
34. Rehman, S.; Al-Abbadi, N.M. Wind shear coefficients and energy yield for Dhahran, Saudi Arabia. *Renew. Energy* **2007**, *32*, 738–749. [CrossRef]
35. Borowy, B.S.; Salameh, Z.M. Optimum Photovoltaic Array Size for a Hybrid Wind/PV System. *IEEE Trans. Energy Convers.* **1994**, *9*, 482–488. [CrossRef]

36. Farrugia, R.N. The wind shear exponent in a Mediterranean island climate. *Renew. Energy* **2003**, *28*, 647–653. [CrossRef]
37. Mohsenian-Rad, A.H.; Wong, V.W.S.; Jatskevich, J.; Schober, R. Optimal and autonomous incentive-based energy consumption scheduling algorithm for smart grid. In Proceedings of the Innovative Smart Grid Technologies Conference, ISGT 2010, Gothenburg, Sweden, 22 March 2010.

Article

Hierarchical Energy Management System for Microgrid Operation Based on Robust Model Predictive Control

Luis Gabriel Marín [1,2,3], Mark Sumner [4], Diego Muñoz-Carpintero [1,5], Daniel Köbrich [1], Seksak Pholboon [4], Doris Sáez [1,6] and Alfredo Núñez [7,*]

1. Department of Electrical Engineering, University of Chile, Santiago 8370451, Chile; luis.marin@ing.uchile.cl (L.G.M.); dimunoz@ing.uchile.cl (D.M.-C.); danielkobrich@gmail.com (D.K.); dsaez@ing.uchile.cl (D.S.)
2. Department of Electrical and Electronics Engineering, Universidad de Los Andes, Bogotá 111711, Colombia
3. Cycle System S.A.S, Bogotá 111311, Colombia
4. Department of Electrical and Electronic Engineering, University of Nottingham, Nottingham NG7 2RD, UK; mark.sumner@nottingham.ac.uk (M.S.); Seksak.Pholboon@nottingham.ac.uk (S.P.)
5. Institute of Engineering Sciences, Universidad de O'Higgins, Rancagua 2841959, Chile
6. Instituto Sistemas Complejos de Ingeniería (ISCI), University of Chile, Santiago 8370397, Chile
7. Section of Railway Engineering, Department of Engineering Structures, Delft University of Technology, 2628CN Delft, The Netherlands

* Correspondence: a.a.nunezvicencio@tudelft.nl

Received: 25 October 2019; Accepted: 20 November 2019; Published: 22 November 2019

Abstract: This paper presents a two-level hierarchical energy management system (EMS) for microgrid operation that is based on a robust model predictive control (MPC) strategy. This EMS focuses on minimizing the cost of the energy drawn from the main grid and increasing self-consumption of local renewable energy resources, and brings benefits to the users of the microgrid as well as the distribution network operator (DNO). The higher level of the EMS comprises a robust MPC controller which optimizes energy usage and defines a power reference that is tracked by the lower-level real-time controller. The proposed EMS addresses the uncertainty of the predictions of the generation and end-user consumption profiles with the use of the robust MPC controller, which considers the optimization over a control policy where the uncertainty of the power predictions can be compensated either by the battery or main grid power consumption. Simulation results using data from a real urban community showed that when compared with an equivalent (non-robust) deterministic EMS (i.e., an EMS based on the same MPC formulation, but without the uncertainty handling), the proposed EMS based on robust MPC achieved reduced energy costs and obtained a more uniform grid power consumption, safer battery operation, and reduced peak loads.

Keywords: hierarchical control; robust control; predictive control; microgrid; uncertainty; prediction interval; energy management system

---

1. Introduction

The integration of large numbers of distributed energy resources (DERs) into the electricity distribution system may play an important role in improving its resilience and sustainability. However, when high penetrations of distributed generation (DG) occur, the management of local and wide-area flow may be compromised and power quality may not satisfy required standards [1].

In [2–4] it is reported that the active management of DG units and controllable loads in different sections of the distribution network (DN) is an acceptable approach for increasing the penetration of DG into a passive DN. The active management of a DN requires the integration of control strategies at different levels in a smart grid framework, as well as communication technologies that allow the connection of DG units to the DN.

This work deals with active management within a DN, namely an energy management system (EMS) for an "energy community". In the context of an increasing trend for small-scale microgrids to encourage the local consumption of energy generated from their RES instead of exporting any surplus to the main grid, energy communities are now appearing where end-user customers manage their local DERs for the benefit of their own microgrids [5]. This may apply to a community that is either geographically co-located or that exists as a virtual entity distributed around a much larger geographical space, with their capabilities "aggregated" by a communications network via web-type services. In this context, the main distribution grid supplies the energy deficit that the microgrid may have. The energy community concept is growing in popularity in the UK, and regulatory change may occur in the foreseeable future that may make the costs of operating this type of community less prohibitive [6].

Hierarchical schemes with multiple levels have been proposed to exploit the benefits of different types of EMS. One possible division lies in the use of optimal controllers (optimal EMSs) or non-optimal controllers (non-optimal EMSs). Most EMSs for scheduling that have been reported in the specialized literature are based on optimal controllers. Loads and energy resources must be predicted in advance, making the effectiveness of optimal approaches dependent mainly on the accuracy of the prediction models. Computation times can also be significantly longer than those for non-optimal EMSs, particularly when using nonlinear predictors. When prediction models cannot capture the behavior of the system or be implemented in real-time, other options are controllers with real-time decision-making capabilities. These can be based on instantaneous power measurements rather than prediction profiles as in [7], or on rules ("rule-based" control) as in [8–11]. For this type of EMS, the aim is usually to reduce energy costs by the efficient use of a battery and maximizing the use of renewable energy to satisfy local demand, while maintaining the reliability of the electrical system. They do not require a detailed model of the system and can respond quickly to changes in the system. However, they are not guaranteed to be optimal and can lead to inefficient energy usage.

Model predictive control (MPC), also known as receding horizon control, is an optimal control strategy that has been used for optimal EMSs. It is based on the optimization of the system's performance over a prediction horizon, which is repeated in each sampling time. Often, the goal of the EMS is to economically manage the DERs to meet certain power quality standards. Therefore, predictions of the renewables and demands are used to find the optimal commitment and dispatch the DER units during a prediction horizon according to some selected performance criteria [12]. Some examples of EMSs based on MPC are reported in [13–18].

An important aspect for optimal EMSs is uncertainty—in this case of the prediction profiles of available renewable energy and end-user consumption [19]. One common paradigm for handling uncertainty is robust optimization, which uses uncertainty sets and combines a worst-case analysis with min–max formulations to obtain optimal solutions that are robust against variations in a parameter with respect to a nominal value (optimal worst-case scenario) [20]. Robust optimization for the scheduling of microgrids has been used for different configurations, such as wind power optimization [21], provisional microgrids [22], and distributed EMSs [23], among others. Robust MPC is a family of MPC controllers which includes robust optimization for handling uncertainties in the predictions, and has also been used for the microgrid EMSs [24–27]. An EMS where the bounds of the uncertainty are given by fuzzy interval models is proposed in [25]. This type of model will be used for the uncertain prediction profiles in this work. All these works dealing with robust optimization find an optimal predicted sequence of control actions that is fixed at each sampling time. However, it is known from the theory of dynamic programming that allowing some compensation of the predicted sequences, as a function of the predicted states or uncertain variables, allows the optimization to find improved solutions. In this case, the optimization is said to be performed over a control policy. Few cases of EMS implement optimization over control policies. While a computationally inefficient (optimization problem with exponentially increasing size with the prediction horizon) robust MPC based EMS is proposed in [28], a more efficient formulation [29] optimizes a predicted sequence of nominal control actions which is corrected by linear terms of disturbances that would affect the system.

In this context, this paper presents a two-level hierarchical EMS for microgrids, where the higher-level controller is based on robust MPC. The aim of the hierarchical two-level architecture, similar to that of [30,31], is to incorporate the benefits of schemes based on both optimal controllers and real-time decision making. Therefore, the EMS comprises a rule-based approach at the lower level with real-time control capabilities and a robust MPC at the higher level to manage the energy efficiency and uncertainties in the predictions of renewable energy resources and end-user load profiles. The main contribution of this work is the design of a robust MPC controller based on fuzzy intervals for the higher level. This controller considers a robust optimization over a control policy parameterized by gains that compensates the uncertainties of the predictions, which are modeled based on fuzzy intervals. The control policy is similar to that of [29], but it was designed according to the particular microgrid considered in this work so that the uncertainty of the power predictions can be compensated either by the battery or main grid power consumption. This compensation enables the controller to find better solutions than other robust MPC formulations with no uncertainty compensation. The predictions of renewable generation and demand are given by fuzzy interval models, which characterize the uncertainty and capture the nonlinearity and temporal dynamics.

Simulation results were obtained using data from a real urban residential community and show the benefits of the proposed strategy. The proposed hierarchical EMS based on robust MPC (robust EMS) achieved a more uniform grid power consumption when compared to the same hierarchical EMS based on MPC (deterministic or non-robust EMS) but without uncertainty handling, since it was able to keep the community power flow closer to the reference power defined by the higher-level controller. It could also achieve safer battery operation and reduced peak loads compared to the deterministic EMS, in addition to typical features of EMSs such as energy cost minimization. The benefits of the robust EMS are due to the incorporation of uncertainty in the formulation and its compensation scheme, which helps the systems to be prepared for errors in the predictions that might yield sub-optimal decisions.

The remainder of this paper is organized as follows: Section 2 presents the problem statement. Section 3 describes the lower level of the EMS: the Community Power Controller at the microgrid level. Section 4 provides the details of the higher level of the EMS: the proposed novel robust predictive control strategy based on fuzzy prediction interval models. Section 5 presents the simulation results showing microgrid operation based on real load and photovoltaic energy profiles from a town in the UK. The last section provides the main conclusions and recommends future work.

2. Problem Statement

A hierarchical EMS as in [30,31] is considered in this work. This paper presents an improvement with respect to these two, as only the lower-level controller within the hierarchy is defined in [30], while uncertainty is not tackled in [31]. The EMS in this work comprises two levels: the microgrid (energy community) level and the main grid level, as shown in Figure 1. Within this framework, the proposed microgrid is composed of domestic demand (a number of non-controllable loads), a number of renewable generation units, and an energy storage system (ESS). Several ESSs could easily be considered in the formulation by including constraints for all of them. However, a single ESS was considered here for simplicity of exposition. This configuration of microgrids is typically associated with groups of dwellings or small villages, and mainly incorporates renewable resources such as photovoltaic arrays and wind generators.

The microgrid can freely use the power from the ESS and the renewable generation, and it can purchase power from the distribution network operator (DNO) for consumption, but it cannot sell. A maximum power limit is set to reduce power peaks of the energy bought from the DNO, and no power can be sent back to it. The ESS also can consume energy in order to store it. The entire renewable generation is either consumed by the loads or stored in the ESS. In this context, the role of the DNO is to supply energy when the renewable generation and the ESS cannot provide enough power to satisfy the demand.

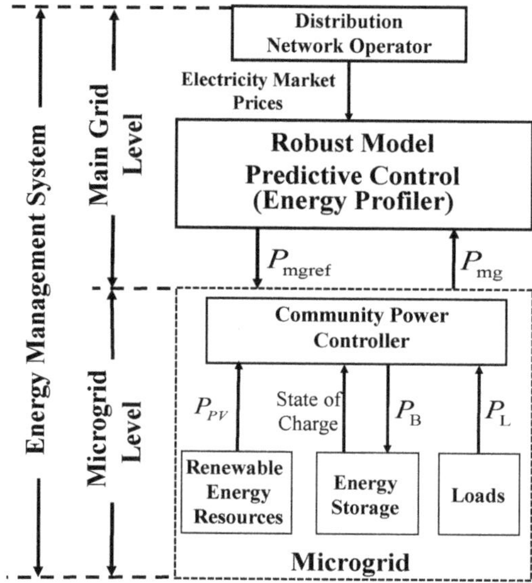

Figure 1. Hierarchical energy management system (EMS) Structure.

At the main grid level, a robust MPC controller operates to provide a realistic power reference (P_{mgref}) for the microgrid, of the power to be consumed from the DNO: this is the "Energy Profiler". At the microgrid level, the "Community Power Controller" aims to track these references in real-time.

The robust MPC implements an optimization of the predicted performance cost given by the price of the energy bought from the main grid, while considering the uncertainty associated with predictions of the renewable generation and consumer load and operational constraints. A sampling time of 30 min was considered because energy markets tend to operate with half-hourly update rates, which defines the update frequency of P_{mgref}.

At the microgrid level, the Community Power Controller operates with a sampling time of 1 min to control the net power flowing from the main grid to the microgrid (P_{mg}) in order to track the power reference (P_{mgref}) sent from the Energy Profiler, while satisfying demand and guaranteeing safe ESS operation.

The following sections present the details of the Community Power Controller at the microgrid level and the robust MPC for the main grid level.

3. Community Power Controller at the Microgrid Level

The ESS is the only dispatchable DER in the proposed microgrid. Thus, the Community Power Controller can only set the charging/discharging power profile (P_B) of the ESS (see Figure 2) in order to track P_{mgref} as sent by the Energy Profiler. $P_B > 0$ indicates the ESS is discharging (generation) and $P_B < 0$ indicates the ESS is charging (load). The ESS consists of a power converter and battery packs, however converter losses are not considered in this study.

The active power of the aggregated microgrid consumption (P_L) and aggregated renewable generation (P_{RG}) are measured at the point of common coupling with a sampling rate of 1 min to calculate the net power (P_{net}) of the microgrid (given by $P_{\mathrm{net}} = P_{\mathrm{L}} - P_{\mathrm{RG}}$). The error between the microgrid power target and the net power is given by:

$$e_{\mathrm{mg}}(k) = P_{\mathrm{mgref}}(k) - P_{\mathrm{net}}(k). \tag{1}$$

Therefore, e_{mg} is the required power from the ESS (P_{B}) so that the instantaneous microgrid power P_{mg} tracks the target P_{mgref} provided by the robust MPC in the Energy Profiler. Based on this, the microgrid-level controller sets the power of the ESS as $P_{\mathrm{B}} = e_{\mathrm{mg}}$ as long as certain constraints are satisfied, as now described.

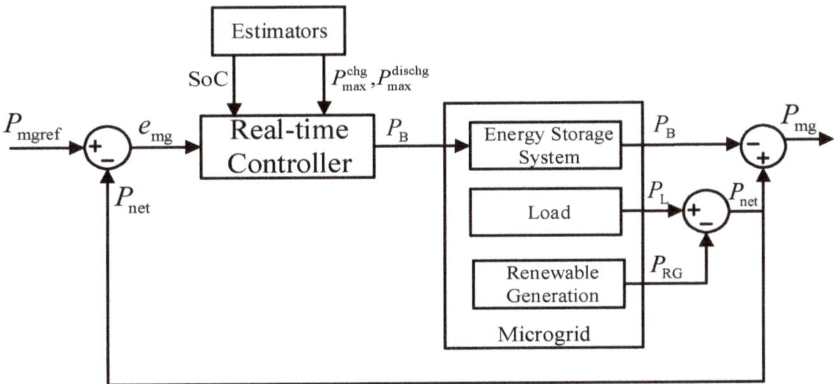

Figure 2. Block diagram at the microgrid level. SoC: state of charge.

For safe operation of the ESS, the maximum available power for charging ($P_{\mathrm{max}}^{\mathrm{chg}}$) and discharging ($P_{\mathrm{max}}^{\mathrm{dischg}}$) is calculated as in [32,33]. These power values are obtained to prevent battery damage by over/under charge (state of charge, SoC) or voltage, or by exceeding the rated current or power limit. The ESS power P_{B} cannot exceed these values. Likewise, $\mathrm{SoC}_{\mathrm{min}} = 0.2$ and $\mathrm{SoC}_{\mathrm{max}} = 0.8$ are the minimum and maximum values allowed for the SoC. These were set to increase the lifespan of the batteries, because capacity fade is typically accelerated by operating profiles with high average SoC and deep discharge levels [34]. To ensure operation within these limits, the SoC value is estimated based on an unscented Kalman filter [35], with outer feedback correction loops as presented in [36]. This is because Bayesian estimation algorithms have been demonstrated to be a well-suited estimation tool for nonlinear problems such as SoC estimation, and they have several advantages including real-time implementation and use of empirical models that better deal with limited and noisy data compared to methods such as ampere-hour counting, internal impedance measurement, and open circuit voltage measurement [37,38].

If the constraints defined above are violated, P_{B} is set as close to e_{mg} as possible to satisfy these constraints. Thus, the microgrid-level controller output P_{B} obeys the following rules:

$$\begin{aligned}
&R1: if\ e_{\mathrm{mg}}(k) \geq 0\ and\ SoC(k) \geq SoC_{\mathrm{max}}\ then\ P_{\mathrm{B}}(k) = 0;\\
&R2: if\ e_{\mathrm{mg}}(k) \geq 0\ and\ SoC(k) < SoC_{\mathrm{max}}\ then\ P_{\mathrm{B}}(k) = -\min(e_{\mathrm{mg}}(k), P_{\mathrm{max}}^{\mathrm{chg}}(k));\\
&R3: if\ e_{\mathrm{mg}}(k) < 0\ and\ SoC(k) \geq SoC_{\mathrm{min}}\ then\ P_{\mathrm{B}}(k) = \min(-e_{\mathrm{mg}}(k), P_{\mathrm{max}}^{\mathrm{dischg}}(k));\\
&R4: if\ e_{\mathrm{mg}}(k) < 0\ and\ SoC(k) < SoC_{\mathrm{min}}\ then\ P_{\mathrm{B}}(k) = -\min(P_{\mathrm{RG}}(k), P_{\mathrm{max}}^{\mathrm{chg}}(k)).
\end{aligned} \tag{2}$$

The instantaneous microgrid power (P_{mg}) is given by:

$$P_{\mathrm{mg}}(k) = P_{\mathrm{net}}(k) - P_{\mathrm{B}}(k), \tag{3}$$

and this tracks P_{mgref} as long as the resulting values of P_{B} and SoC do not violate constraints.

4. Robust Model Predictive Control

The role of the higher-level controller is to calculate the reference power (P_{mgref}) so that it minimizes the energy cost for the community, but also ensures that it can be tracked reasonably well by the Community Power Controller based on the available resources (P_{B} and P_{RG}) and load (P_{L}).

The proposed EMS is based on robust MPC, and thus it requires models to predict the expected value and variability of the demand, as well as the energy available from the renewable resources over a prediction horizon. Clearly, the performance of the robust EMS depends on the quality of these models. In this work, fuzzy prediction interval models are used, as presented next.

4.1. Fuzzy Prediction Interval Model

Fuzzy prediction interval models are used to predict the expected values and the uncertainty of the net power of the microgrid (P_{net}). These predictions are used in the main grid-level robust MPC, with a sampling time of 30 min. Since the original data has a 1 min resolution, $P_{\text{net}}(k)$ represents the average of the measurements (made once per minute) for the 30 min following time instant k.

The fuzzy prediction interval model proposed in [39] is adopted in this work. The fuzzy model for obtaining the predicted expected value of P_{net} is given by

$$\hat{P}_{\text{net}}(k) = \sum_{r=1}^{R} \beta^r(Z(k))\hat{P}^r_{\text{net}}(k) = \sum_{r=1}^{R} \beta^R(Z(k))[1\ Z(k)]\theta_r = \Psi^T\Theta, \tag{4}$$

where $Z(k) = [P_{\text{net}}(k-1), \ldots, P_{\text{net}}(k-N_y)]$, the number of rules is R, β^r is the activation degree, θ_r is the coefficient vector of the consequences, and $\hat{P}^r_{\text{net}}(k) = [1\ Z(k)]\theta_r$ is the local output at time k of rule r, with $r = 1, \ldots, R$. $\Psi^T = [\psi_1^T, \ldots, \psi_R^T]$ is the fuzzy regression matrix, and $\Theta^T = [\theta_1^T, \ldots, \theta_R^T]$ is the coefficient matrix for all rules. The maximum regressor order corresponds to one day before ($N_y = 48$), and some of these input variables can be discarded using a sensitivity analysis [40]. The Gustafson–Kessel clustering algorithm is used to find R and the parameters of $\beta^r(\cdot)$. Parameters Θ are estimated by the least-squares method [41].

The predictions for j steps ahead made at time k are:

$$\hat{P}_{\text{net}}(k+j) = \sum_{r=1}^{R} \beta^r(Z(k+j))\hat{P}^r_{\text{net}}(k+j), \tag{5}$$

where $Z(k+j) = [P_{\text{net}}(k+j-1), \ldots, P_{\text{net}}(k+j-N_y)], j = 1, \ldots, N$.

Fuzzy prediction interval models provide the lower ($\underline{\hat{P}}_{\text{net}}(k+j)$) and upper ($\overline{\hat{P}}_{\text{net}}(k+j)$) bounds predicted at time k such that the real values of $P_{\text{net}}(k+j)$ satisfy $\underline{\hat{P}}_{\text{net}}(k+j) \leq P_{\text{net}}(k+j) \leq \overline{\hat{P}}_{\text{net}}(k+j)$, with a certain coverage probability p, for $j = 1, \ldots, N$ where N is the prediction horizon. It is proposed in [39] that the lower and upper bounds ($\underline{\hat{P}}_{\text{net}}$) and ($\overline{\hat{P}}_{\text{net}}$) are estimated by

$$\overline{\hat{P}}_{\text{net}}(k+j) = \hat{P}_{\text{net}}(k+j) + \alpha_{k+j} I^{\text{TS}}(k+j), \tag{6}$$

$$\underline{\hat{P}}_{\text{net}}(k+j) = \hat{P}_{\text{net}}(k+j) - \alpha_{k+j} I^{\text{TS}}(k+j), \tag{7}$$

where $I^{\text{TS}}(k+j) = \sum_{r=1}^{R} \beta^r(Z^*(k+j)) I_r^{\text{TS}}(k+j)$, with $I_r^{\text{TS}}(k+j) = \hat{\sigma}_r \left(1 + \psi_r^{*T}\left(\psi_r\psi_r^T\right)^{-1}\psi_r^*\right)^{1/2}$, is the component associated with the covariance of the error between the observed data and the local model outputs. The current input ψ_r^{*T} is associated to a new datum $Z^*(k+j)$. Additionally, α_{k+j} are scaling parameters that are tuned using experimental data so that the interval defined by $[\underline{\hat{P}}_{\text{net}}(k+j), \overline{\hat{P}}_{\text{net}}(k+j)]$ contains the actual values of $P_{\text{net}}(k+j)$ with a given coverage probability. The next section presents deterministic and robust MPC formulations using fuzzy prediction interval modeling.

The deterministic MPC is presented to illustrate the basics of the formulation, and it will be used as a basis for comparison. Focus is later given to the robust MPC, which is the main contribution of this work.

4.2. Deterministic EMS

The role of the MPC scheme at the main-grid level is to minimize the cost of the power delivered to the microgrid from the main grid. This controller uses a model of the microgrid dynamics to find its predicted behavior, which assumes that there are no losses, nor congestion or voltage regulation issues for the power transfer from the DNO to the microgrid and between elements within the microgrid. The sampling time of the model and the controller is $T_s = 30$ min. The prediction horizon of the controller is $N = 48$; the power references one day ahead (48 steps with $T_s = 30$ min) are found to optimize the predicted behavior for a one-day ahead operation. More precisely, at each discrete time k, an optimization problem uses this model to find the optimal sequence of $P_{\text{mgref}}(k+j), j = 0, \ldots, N-1$ that minimizes the energy consumption during the prediction horizon N.

The system dynamic is given by the evolution of the energy in the ESS (E_{B}). These dynamics must be included in the MPC optimization, and are described by a simplified linear model:

$$E_{\text{B}}(k+j+1) = E_{\text{B}}(k+j) - T_s P_{\text{B}}(k+j). \tag{8}$$

The prediction of future states requires an estimation of the current state, obtained from the unscented Kalman filter at the microgrid level, which sends this information to the upper layer.

The power balance at the microgrid level must also be imposed in the MPC optimization. This constraint is invoked as

$$P_{\text{mgref}}(k+j) = \hat{P}_{\text{net}}(k+j) - P_{\text{B}}(k+j). \tag{9}$$

Here, the net power of the microgrid is given by its expected values $\hat{P}_{\text{net}}(k+j)$, which are obtained by the fuzzy prediction model defined in (5). Other constraints that must be considered in the optimization include the minimum and maximum limits of battery capacity:

$$E_{\text{min}} = 0.2C_n \le E_{\text{B}}(k+j) \le E_{\text{max}} = 0.8C_n, \tag{10}$$

where C_n is the nominal capacity, and the limits for charging and discharging of the ESS are

$$-P_{\text{max}}^{\text{dischg}}(k+j) \le P_{\text{B}}(k+j) \le P_{\text{max}}^{\text{chg}}(k+j), \tag{11}$$

where the bounds are approximated linearly, such that $P_{\text{max}}^{\text{dischg}}(k+j) = \alpha_d P_{\text{B}}^{\text{max}} SoC(k+j)$ and $P_{\text{max}}^{\text{chg}}(k+j) = \alpha_c P_{\text{B}}^{\text{max}}(1 - SoC(k+j))$. Here, $P_{\text{B}}^{\text{max}}$ is the maximum instantaneous power given by the manufacturer, and α_d and α_c are tuned parameters which avoid violation of the under/over SoC limits, respectively. The last constraints to be used are the minimum and maximum grid powers:

$$-P_{\text{mg}}^{\text{min}} \le P_{\text{mgref}}(k+j) \le P_{\text{mg}}^{\text{max}}. \tag{12}$$

Since the EMS aims to maximize self-consumption (i.e., minimize energy exported to the main grid) and to minimize the power drawn from the main grid during peak periods, $P_{\text{mg}}^{\text{min}} = 0$ was chosen, and $P_{\text{mg}}^{\text{max}}$ can be chosen arbitrarily in order to reduce power peaks that are purchased from the main grid.

With these considerations, and because the EMS also aims to minimize costs, the optimal control problem to be solved at time k is given by:

$$\min_{\{P_{\text{mgref}}(k+j)\}_{j=0,\dots,N-1}} \sum_{j=0}^{N-1} C(k+j)P_{\text{mgref}}(k+j)T_s \tag{13}$$
$$\text{subject to (8)–(12) all for } j = 0,\dots,N-1,$$

where $C(k+j)$ is the energy price which is a known parameter for the EMS and is based on a Time of Use tariff scheme; the price of the unit of energy depends on the hour within the day. This is a linear program. In this paper, we solved the problem using a Matlab implementation of an interior-point algorithm for linear programs. Finally, only the first element of the sequence $\{P_{\text{mgref}}(k+j)\}_{j=0,\dots,N-1}$ (namely, $P_{\text{mgref}}(k)$) is actually sent as a reference to the microgrid, and the procedure is repeated at time $k+1$ (i.e., 30 min ahead).

4.3. Robust EMS with Explicit Uncertainty Compensation

The formulation of Section 4.2 ignores the uncertainty of the predictions of P_{net}. While the closed-loop nature of the controller provides some robustness to uncertainty, its explicit inclusion in the formulation may bring further benefits in performance, as discussed in [42], and will be seen in Section 5. This section deals with uncertainty handling in the controller formulation.

Fuzzy prediction interval models are used to model the uncertainty of P_{net} predictions. The real values $P_{\text{net}}(k+j)$ satisfy $P_{\text{net}}(k+j) = \hat{P}_{\text{net}}(k+j) + \Delta P_{\text{net}}(k+j)$, where $\hat{P}_{\text{net}}(k+j)$ is the expected value of the prediction and $\Delta P_{\text{net}}(k+j)$ is the deviation of the actual value from the prediction. This deviation is uncertain, but satisfies

$$\Delta P_{\text{net}}(k+j) \in [\Delta \hat{P}^{\min}_{\text{net}}(k+j), \Delta \hat{P}^{\max}_{\text{net}}(k+j)], \tag{14}$$

where

$$\begin{aligned} \Delta \hat{P}^{\max}_{\text{net}}(k+j) &= \overline{\hat{P}}_{\text{net}}(k+j) - \hat{P}_{\text{net}}(k+j) \\ \Delta \hat{P}^{\min}_{\text{net}}(k+j) &= \underline{\hat{P}}_{\text{net}}(k+j) - \hat{P}_{\text{net}}(k+j), \end{aligned} \tag{15}$$

for $j = 1,\dots,N-1$. These intervals are designed to have a minimum interval width and guarantee that the future real values fall within the interval with a certain coverage probability.

The solution for deterministic optimal control problems such as deterministic MPC is a sequence of fixed control actions. However, this is conservative when there are uncertain components, as this ignores the fact that there will be a correction of the disturbances by the closed-loop operation of the controller. Instead, finding a sequence of control actions or decision variables that depend on the predicted states or that can be corrected with the predicted values of uncertain variables allows the optimization to find improved solutions. It is shown in [43,44] that a computationally efficient alternative to acknowledge these corrections in the optimization is to explicitly compensate the uncertain terms with linear gains $L(k+j)$. The robust MPC formulation proposed here follows this idea, but was adapted to satisfy the power balance constraint (9). As a result, the compensation is performed either by the ESS or the main grid consumption.

The following control laws for the predicted inputs of the optimization at time k, P_B and P_{mgref}, which are coupled by (9), are proposed:

$$P_B(k+j) = \hat{P}_B(k+j) + L(k+j)\Delta \hat{P}_{\text{net}}(k+j), \tag{16}$$

$$P_{\text{mgref}}(k+j) = \hat{P}_{\text{mgref}}(k+j) + (1 - L(k+j))\Delta \hat{P}_{\text{net}}(k+j), \tag{17}$$

where $\hat{P}_{\text{mgref}}(k+j)$, $\hat{P}_B(k+j)$ and $L(k+j)$ are the optimization variables for $j = 0,\dots,N-1$. This can be interpreted as follows: if $P_{\text{net}}(k+j) = \hat{P}_{\text{net}}(k+j)$ (thus $\Delta \hat{P}_{\text{net}}(k+j) = 0$), then $P_B(k+j) = \hat{P}_B(k+j)$.

Otherwise, the predicted input to be applied to the system is compensated by $L(k+j)\Delta\hat{P}_{\text{net}}(k+j)$. Note that the compensation $P_{\text{mgref}}(k+j)$ is given by $(1-L(k+j))\Delta\hat{P}_{\text{net}}(k+j)$, so that the balance equation for the expected values

$$\hat{P}_{\text{mgref}}(k+j) = \hat{P}_{\text{net}}(k+j) - \hat{P}_{\text{B}}(k+j) \tag{18}$$

is enough to satisfy the full balance for the real values (9). The following constraint is used on $L(k+j)$:

$$0 \leq L(k+j) \leq 1, \tag{19}$$

which indicates that the deviation of the real value from the prediction is compensated by $P_{\text{B}}(k+j)$ and $P_{\text{mgref}}(k+j)$ in a proportion defined by $L(k+j)$.

The predicted control laws of (16) and (17) depend on the uncertain values $\Delta\hat{P}_{\text{net}}(k+j)$, and so will the predictions of E_{B}. However, the optimization problem as posed in (13) (a linear program) cannot be solved with uncertain values. A worst-case approach is taken, where $\Delta\hat{P}_{\text{net}}(k+j)$ are assigned to take the worst possible values according to some criterion. Consider (12), which imposes the limits for P_{mgref} and in the current setting is equivalent to

$$\begin{aligned}\hat{P}_{\text{mgref}}(k+j) + (1-L(k+j))\Delta\hat{P}_{\text{net}}(k+j) &\leq P_{\text{mg}}^{\text{max}},\\ -\hat{P}_{\text{mgref}}(k+j) - (1-L(k+j))\Delta\hat{P}_{\text{net}}(k+j) &\leq P_{\text{mg}}^{\text{min}}.\end{aligned}$$

These inequalities depend on $\Delta\hat{P}_{\text{net}}(k+j)$, which is uncertain, so it is not known what value it will take. Therefore, these are enforced by taking a worst-case approach, as is common in robust MPC. They are implemented by setting $\Delta\hat{P}_{\text{net}}(k+j)$ to take the values that reduce freedom the most for $\hat{P}_{\text{mgref}}(k+j)$ in each of the inequalities: $\Delta\hat{P}_{\text{net}}^{\text{max}}(k+j)$ and $\hat{P}_{\text{net}}^{\text{min}}(k+j)$, respectively. Thus, the constraints above are enforced in the optimization as

$$\begin{aligned}\hat{P}_{\text{mgref}}(k+j) + (1-L(k+j))\Delta\hat{P}_{\text{net}}^{\text{max}}(k+j) &\leq P_{\text{mg}}^{\text{max}},\\ \hat{P}_{\text{mgref}}(k+j) + (1-L(k+j))\Delta\hat{P}_{\text{net}}^{\text{min}}(k+j) &\geq P_{\text{mg}}^{\text{min}}.\end{aligned} \tag{20}$$

Note that these constraints are linear because $\Delta\hat{P}_{\text{net}}^{\text{max}}(k+j), \Delta\hat{P}_{\text{net}}^{\text{min}}(k+j)$ are known for the optimization, and only $\hat{P}_{\text{mgref}}(k+j)$ and $L(k+j)$ are optimization variables. For all constraints associated with the ESS, the worst case is considered to be that where $P_{\text{net}}(k+j)$ is the largest; that is, $\Delta P_{\text{net}}(k+j) := \Delta\hat{P}_{\text{net}}^{\text{max}}(k+j)$. This is the case with the most deficit of renewables with respect to demand, which is the instant where the ESS is needed the most to provide flexibility and reduce the energy bought from the grid. Therefore, constraints (8), (10), and (11) are reformulated as:

$$-T_s\sum_{i=0}^{j}\hat{P}_{\text{B}}(k+i) - T_s\sum_{i=0}^{j}L(k+i)\Delta\hat{P}_{\text{net}}^{\text{max}}(k+i) \leq E_{\text{max}} - E_{\text{B}}(k), \tag{21}$$

$$T_s\sum_{i=0}^{j}\hat{P}_{\text{B}}(k+i) + T_s\sum_{i=0}^{j}L(k+i)\Delta\hat{P}_{\text{net}}^{\text{max}}(k+i) \leq -E_{\text{min}} + E_{\text{B}}(k), \tag{22}$$

$$\begin{aligned}\hat{P}_{\text{B}}(k+j) + L(k+j)\Delta\hat{P}_{\text{net}}^{\text{max}}(k+j) &\leq P_{\text{max}}^{\text{dischg}}(k+j),\\ \hat{P}_{\text{B}}(k+j) + L(k+j)\Delta\hat{P}_{\text{net}}^{\text{max}}(k+j) &\geq P_{\text{max}}^{\text{chg}}(k+j).\end{aligned} \tag{23}$$

With all these considerations, the optimization problem to be solved at each time k is

$$\min_{x} \sum_{j=0}^{N-1} C(k+j)\hat{P}_{\text{mgref}}(k+j)\ T_s \tag{24}$$
$$\text{subject to (18)–(23) all for } j = 0, \ldots, N-1,$$

where $x = \{\widehat{P}_{\text{mgref}}(k+j), \widehat{P}_{\text{B}}(k+j), L(k+j)\}_{j=0,\ldots,N-1}$.

This is also a linear program, and is solved with the same Matlab solver as for (13). Finally, $P_{\text{mgref}}(k)$ is sent as a reference to the microgrid, and the procedure is repeated at time $k+1$.

Using this robust MPC guarantees the satisfaction of constraints for the worst cases incorporated in the optimization. For instance, it ensures that the power reference sent does not instruct the lower level to sell energy to the grid nor that the power bought is greater than the upper limit. On the other hand, using worst-case constraints may introduce conservativeness to the solutions, which may be reflected as economic costs, because worst cases may not occur.

5. Case Study

The performance of the hierarchical EMS based on robust MPC was tested by the simulation of a community connected to the main grid, made up of 30 dwellings with a 50% level of photovoltaic power penetration (i.e., 15 dwellings have a photovoltaic array) and an ESS made of lead-acid batteries with a 135-kWh capacity.

Data for winter from a town in the UK was used [45]. For this scenario, a three-level Time of Use tariff (similar to [46]) was considered for buying energy from the grid for each day of the simulation. The prices are shown in Table 1.

Table 1. Energy price during the day.

Hours	00:00–06:00	06:00–16:00	16:00–19:00	19:00–23:00	23:00–24:00
Energy Cost	5 p/kWh	12 p/kWh	25 p/kWh	12 p/kWh	5 p/kWh

5.1. Fuzzy Prediction Interval for Net Power of the Microgrid

Load and photovoltaic power data available for a town in the UK were used to develop the fuzzy prediction interval model described in Section 4.1 for the net power given by $P_{\text{net}} = P_{\text{L}} - P_{\text{RG}}$. The data cover a period of 90 days corresponding to the winter season, and this was divided into training, validation, and test data sets. The maximum value of P_{net} was 67.57 kW and the minimum value was −45.09 kW, and a sampling time of 30 min was used.

The fuzzy model and regressors obtained during the identification for the predictor of $P_{\text{net}}(k)$ were:

$$\hat{P}_{\text{net}}(k) = f^{\text{fuzzy}}(P_{\text{net}}(k-i_1), \ldots, P_{\text{net}}(k-i_n)), \tag{25}$$

where $\{i_1, \ldots, i_n\} = \{1, 2, 8, 25, 26, 32, 38, 42, 43, 44, 46, 48\}$ and the optimal structure of the model has four rules. Note that exogenous variables were not included in the model.

The prediction interval coverage probability (PICP), which quantifies the proportion of measured values that fall within the predicted interval, and the prediction interval normalized average width (PINAW), which quantifies the width of the interval, were used as indexes to evaluate the quality of the interval for h-step-ahead predictions:

$$\text{PICP}(h) = \frac{1}{T}\sum_{k=1}^{T} \delta_{k+h}, \tag{26}$$

$$\mathrm{PINAW}(h) = \frac{1}{TR}\sum_{k=1}^{T}\left(\overline{\hat{P}}_{\mathrm{net}}(k+h) - \underline{\hat{P}}_{\mathrm{net}}(k+h)\right), \tag{27}$$

for $h = 1,\ldots,N$, where $P_{\mathrm{net}}(k)$ is the real value of P_{net}, R is the distance between the maximum and minimum values of $P_{\mathrm{net}}(k)$ in the data set, and $\delta_{k+h} = 1$ if $P_{\mathrm{net}}(k+h) \in [\overline{\hat{P}}_{\mathrm{net}}(k+h), \underline{\hat{P}}_{\mathrm{net}}(k+h)]$; otherwise, $\delta_{k+h} = 0$. Additionally, the root mean square error (RMSE) and the mean absolute error (MAE) were used to evaluate the accuracy of the prediction model associated with the expected value.

In this study, the prediction interval model was tuned at a PICP of 90% for all prediction instants. Table 2 shows the performance indexes associated with three different prediction horizons for the test dataset. The results indicate that the fuzzy prediction interval was effectively tuned to a PICP of 90%, and that the interval width (PINAW) increased with the prediction horizon. Figure 3 shows the one-day-ahead prediction intervals for three days of the test dataset. The red line is the one-ahead prediction $(\hat{P}_{\mathrm{net}})$ of the net power of the microgrid (P_{net}), the blue points are the actual data (P_{net}) used to evaluate the performance of the fuzzy prediction interval model, and the grey box is the prediction interval which is characterized by the lower $(\underline{\hat{P}}_{\mathrm{net}})$ and upper $(\overline{\hat{P}}_{\mathrm{net}})$ bounds.

The expected value $(\hat{P}_{\mathrm{net}})$ and lower $(\underline{\hat{P}}_{\mathrm{net}})$ and upper $(\overline{\hat{P}}_{\mathrm{net}})$ bound predictions provided by the prediction interval were used in the deterministic and robust EMSs, as explained in Sections 4.2 and 4.3.

Table 2. Performance indices of fuzzy prediction interval model. MAE: mean absolute error; PICP: prediction interval coverage probability; PINAW: prediction interval normalized average width; RMSE: root mean square error.

Performance Indices	Prediction Horizon		
	One Hour Ahead	Six Hours Ahead	One Day Ahead
RMSE (kW)	4.5136	5.0471	5.1974
MAE (kW)	3.2995	3.7316	3.7530
PINAW (%)	22.73	27.62	28.02
PICP (%)	88.22	89.79	89.83

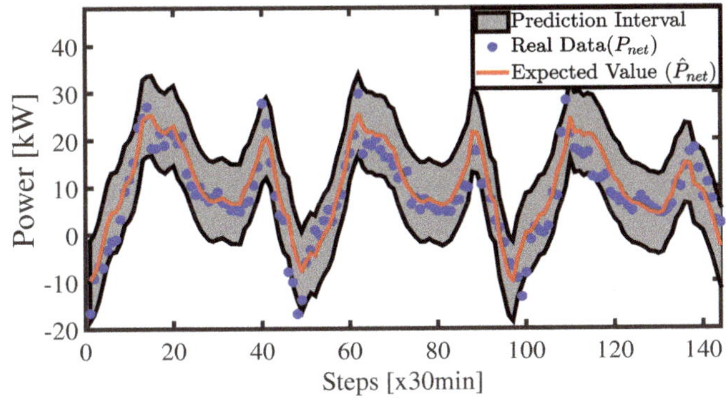

Figure 3. One-day-ahead prediction interval for P_{net} tuned at $PICP = 90\%$.

5.2. Hierarchical EMS Results

The performance of the EMS based on robust MPC with fuzzy interval models (Section 4.3) is analyzed in this section. For this purpose, it was compared with the deterministic EMS presented in Section 4.2. Simulation results for this comparison are presented in the following.

Figure 4 shows the responses obtained with the hierarchical EMSs (deterministic and robust) for operation over two days. Results were consistent with the daily distribution of the energy prices.

Since energy from the main grid was most expensive in the 16:00–19:00 h time block, the EMS controlled this power to be close to zero. The opposite behavior occurred during morning and late night hours (0:00–06:00 and 23:00–24:00) when the energy price was considerably cheaper. It can also be seen that in both deterministic and robust approaches the power reference (P_{mgref}), as sent by the higher-level MPC controller (in red), could be tracked reasonably well by the lower-level controller (P_{mg}, in blue). Tracking errors occurred when the maximum available battery power for charging or discharging was less than the ESS power required by the microgrid (see the rules in Section 3). Additionally, Figure 4 shows that the robust EMS found a flatter P_{mgref} than the deterministic EMS, which is good for the distribution network operator because it minimizes the grid power profile fluctuations. Several metrics justify and quantify the flattening, as will be discussed below.

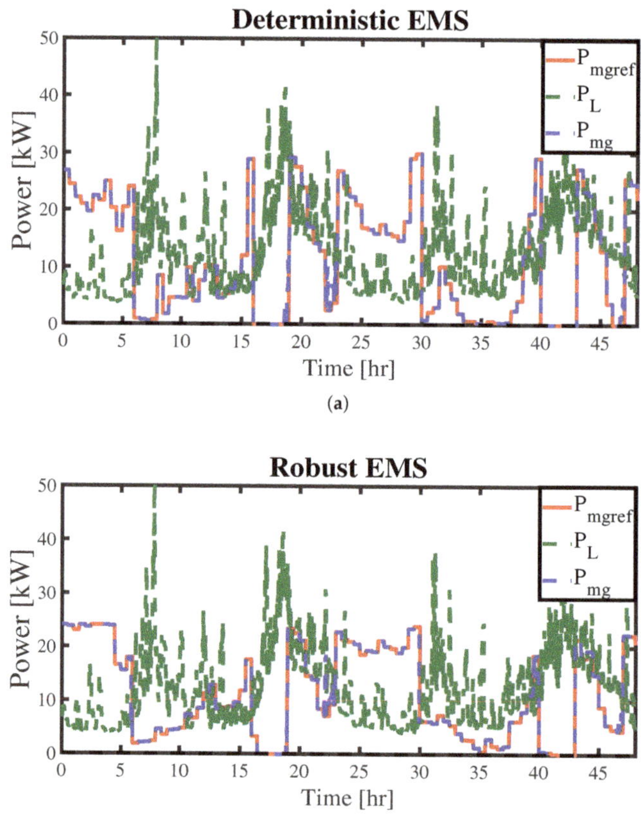

Figure 4. Performance of the proposed hierarchical EMS: (**a**) Deterministic approach; (**b**) Robust approach.

Table 3 shows the energy costs, the RMSE of the tracking error of the power reference (P_{mgref}), the equivalent full cycles (EFC), and the loss of power supply probability (LPSP) for one week of simulation using the deterministic and robust EMSs (see Appendix A for definition of EFC and LPSP). It can be seen that the robust EMS reached a better operation cost than the deterministic EMS. Additionally, the lower RMSE with the robust EMS means that there was a better tracking of the power reference (P_{mgref}) sent by the higher level to the microgrid (see Figure 4). The lower EFC of the robust EMS means that fewer cycles were used by the ESS which directly improved the state-of-health and lifetime of the ESS. As battery aging (measured by the state-of-health) is a function of the elapsed time from the manufacture date, as well as the usage by consecutive charge and discharge actions, a lower EFC improves the battery life time. Finally, the LPSP, which is the fraction of time where the microgrid cannot fulfill the load requirements using the reference power (P_{mgref}) defined by the higher level and the available resources of the microgrid (renewable generation and ESS), was 3.780% for the deterministic EMS and 2.927% for the robust EMS. This was because the robust approach compensated for the uncertainty of generation and demand and could avoid the scenarios measured by the LPSP.

Table 3. Performance indices during a simulation of one-week duration. EFC: equivalent full cycles; LPSP: loss of power supply probability.

EMS Strategy	Cost	RMSE	EFC	LPSP
	(£)	(kW)	Cycles	(%)
Deterministic EMS	168.01	1.22	6.40	3.780
Robust EMS	165.28	1.14	6.07	2.927

Table 4 shows the energy bought by the community from the main grid during the time periods associated with different tariff prices. C1 is the time with the cheapest price and C3 is the time with the highest price. As discussed above, the operation of both hierarchical EMSs was consistent with these price bands: more energy was bought at C1 and C2, less energy was bought at C3. Note that the robust EMS bought more at C1 than the deterministic EMS. However, it spent less in C2 and considerably less than the deterministic EMS at C3. It is apparent then that the robust EMS managed to obtain savings with respect to the deterministic EMS by being better at planning against worst cases; namely, it avoided buying energy when it was most expensive.

Table 4. Energy distribution at different prices.

EMS Strategy	C1	C2	C3
	(kWh)	(kWh)	(kWh)
Deterministic	990.361	934.338	25.483
Robust	994.081	931.231	15.321

Finally, for further evaluation of the EMSs, several indexes of operation are presented in Table 5. These are the load factor (LF), the load loss factor (LLF), positive power peak (P^{+}), negative power peak (P^{-}), the maximum power derivative (MPD), and the average power derivative (APD). See Appendix A for detailed definitions, but the interpretations of these are presented next.

Table 5. Quality indexes for the power profile of the main grid. APD: average power derivative; LF: load factor; LLF: load loss factor; MPD: maximum power derivative.

EMS Strategy	LF	LLF	P^{+}	P^{-}	MPD	APD
			(kW)	(kW)	(kW/min)	(kW/min)
Deterministic	0.3869	0.2452	30.00	0	29.63	0.1889
Robust	0.4459	0.2880	25.90	0	22.48	0.1318

The LF describes the flatness of the power response: values close to 1 are associated with flat responses while values close to 0 indicate the presence of large peaks. The LLF quantifies the losses incurred as a result of peak power: values close to 1 describe flat responses with small losses, while values close to 0 indicate large losses due to large peaks [7]. The MPD is the maximum value of the rate of change between two consecutive points of the main grid power in its absolute value [10,47]. The APD is the average of the absolute value of the rate of change of the main grid power.

The LF was greater for the robust EMS than for the deterministic case (LF = 0.4459 and LF = 0.3869, respectively). This clearly indicates that the response for the robust EMS was flatter (which is also consistent with the results of RMSE and EFC reported above). Similarly, LLF = 0.288 for the robust EMS, and LF = 0.2452 for the deterministic EMS. Therefore, the hierarchical EMSs resulted in a reduction of the peak power and a reduction of losses due to peak power.

The positive power peak (P^+) and negative power peak (P^-) for the hierarchical EMS were limited by constraints as explained in Section 4. The limits were $P_{\text{mg}}^{\text{min}} = 0$ kW, which guarantees that no energy was exported to the main grid, and $P_{\text{mg}}^{\text{max}} = 30$ kW. The robust EMS works in a more conservative manner for the upper limit. It attempts to avoid sub-optimal operation due to worst-case scenarios: thus, it allows smaller peaks $(P^+ = 25.9)$ kW than the deterministic EMS (30 kW) (see also Figure 5).

The last two metrics were also improved using the proposed robust-MPC-based EMS: the MPD and APD were reduced compared with the deterministic EMS. Finally, a lower APD corresponds to a flatter main grid power, which is consistent with previously analyzed indicators.

Overall, it can be seen that the deterministic and robust hierarchical EMSs provide mechanisms for efficient energy management. However, the robust EMS provided improvements over the deterministic EMS, which can be explained because the uncertainty management in the robust EMS helps the system to be prepared for errors in the predictions that might yield sub-optimal decisions.

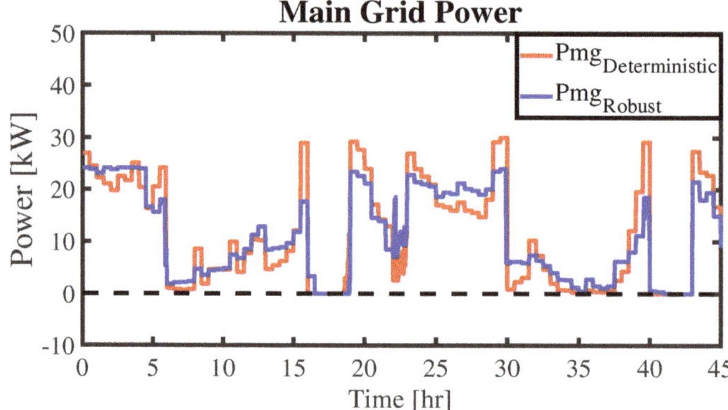

Figure 5. Main grid power profiles.

6. Conclusions

In this paper, a two-level hierarchical EMS based on robust MPC was presented for the operation of energy communities (microgrids), considering the uncertainty of the renewable energy resources and electrical load consumption. The robust MPC has a special structure which enables compensation of the uncertain predictions by the battery within the microgrid or consumption from the main grid.

While the deterministic EMS could effectively operate the microgrid, the robust EMS consistently performed better over several indicators of performance considered in the work. Most importantly: improved operational cost, flatter response of the power drawn from the main grid, and a greater capacity to satisfy demand from the microgrid. This is because robust MPC handles uncertainty and prepares better for unexpected changes in the microgrid generation or loads.

Future work will incorporate real-time prices in the formulation to reflect the price on the wholesale market. A market scheme that allows the selling of excess energy from the microgrid to the main grid will also be considered. Additionally, the benefits of the proposed hierarchical EMS based on robust MPC will be explored for energy communities such as factories, schools, commercial parks, among others; other communities using different types of loads; distributed generations, such as biomass-based generation; or energy storage technologies, such as hydrogen or flywheels. Finally, the incorporation of demand-side management (DSM) strategies into the robust

MPC formulation could be studied to determine an optimal demand schedule, helping to generate desired changes in the load profile.

Author Contributions: Conceptualization, L.G.M., M.S., S.P., and D.S.; methodology, L.G.M., D.M.-C., and D.S.; software, L.G.M. and D.K.; validation, L.G.M., D.M.-C., and D.S.; formal analysis, L.G.M., D.M.-C., and A.N.; Writing—Original draft preparation, L.G.M.; Writing—review and editing, L.G.M., M.S., D.M.-C., D.K., S.P., D.S., and A.N.

Funding: This research was funded by the Instituto Sistemas Complejos de Ingeniería (ISCI) (CONICYT PIA/BASAL AFB180003), the Solar Energy Research Center SERC-Chile (CONICYT/FONDAP/Project under Grant 15110019), FONDECYT Chile Grant Nr. 1170683 "Robust Distributed Predictive Control Strategies for the Coordination of Hybrid AC and DC Microgrids" and CONICYT-FONDECYT Postdoctorado-3170040. Luis Gabriel Marín has also been supported by a Ph.D. scholarship from COLCIENCIAS-Colombia and by a scholarship from CONICYT-PCHA/Doctorado Nacional para extranjeros/2014-63140093.

Conflicts of Interest: The authors declare no conflicts of interest.

Abbreviations

The following abbreviations are used in this manuscript:

EMS	Energy management system
MPC	Model predictive control
DNO	Distribution network operator
DER	Distributed energy resource
DG	Distributed generation
DN	Distribution network
ESS	Energy storage system
SoC	State of charge
PICP	Prediction interval coverage probability
PINAW	Prediction interval normalized average width
RMSE	Root mean square error
MAE	Mean absolute error
EFC	Equivalent full cycles
LPSP	Loss of power supply probability
LF	Load Factor
LLF	Load loss factor
MPD	Maximum power derivative
APD	Average power derivative

Appendix A. Performance Indices for the Power Profile of the Main Grid

Several indexes that evaluate the quality of the power profiles sent from the main grid to the micro-grid were used to compare the results obtained with the different control strategies. These indices are described in the following.

RMSE is the root mean square error:

$$\text{RMSE} = \sqrt{\frac{\sum_{k=1}^{T} \left(P_{\text{mgref}}(k) - P_{\text{mg}}(k)\right)^2}{T}}, \tag{A1}$$

which represents the capability of the microgrid to follow the power reference (P_{mgref}) sent by the higher level in the hierarchical EMS.

The equivalent full cycles (EFC) is the number of full discharges that an ESS performs throughout its time use [48]:

$$\text{EFC} = \frac{E_{\text{dis}}(Ah)}{C_{\text{n}}}, \tag{A2}$$

where $E_{\text{dis}}[Ah]$ is the discharge energy during the simulation time and C_{n} is the nominal battery capacity. The EFC is a metric associated with the life cycle of the ESS. In this approach, one cycle per day is the desired EFC ($\text{EFC}_{\text{desired}}$).

The loss of power supply probability (LPSP) is the ratio between the energy deficiency and the total energy demands for a period of time [49]. In this approach, the energy deficiency occurs when $(P_{\text{net}}(k) - P_{\text{mgref}}(k))T_{\text{s}} > 0$, which means that the available maximum power of the ESS ($P_{\text{max}}^{\text{dischg}}$) cannot fulfill the load, and therefore the energy deficiency is supplied from the main grid. When this happens, the microgrid cannot follow the power reference (P_{mgref}) perfectly, and therefore $P_{\text{mg}} = P_{\text{mgref}} + \text{ED}$. A lower value of LPSP indicates a higher probability that the load will be satisfied. The LPSP is defined as [50]

$$\text{LPSP} = \frac{\sum_{k=1}^{T} T_k}{T}, \tag{A3}$$

where T_k is the number of instants when an energy deficiency occurs and T is the total simulation time. The load factor (LF) is given by

$$\text{LF} = \frac{\text{Avg}(P_{\text{mg}})}{\max(P_{\text{mg}})}, \tag{A4}$$

and quantifies the ratio between the average grid power ($P_{\text{mg}}^{\text{AVG}}$) and peak grid power ($P_{\text{mg}}^{\text{max}}$) during a given period. An improvement to the LF value indicates a peak load reduction.

The load loss factor (LLF) is a measure of losses incurred as a result of peak power:

$$\text{LLF} = \frac{\text{Avg}(P_{\text{mg}}^2)}{\max(P_{\text{mg}}^2)}. \tag{A5}$$

The maximum power derivative (MPD) is the maximum value of the rate of change between two consecutive points of the main grid power in its absolute value:

$$\text{MPD} = \max(|\Delta P_{\text{mg}}(k)|), \tag{A6}$$

where $\Delta P_{\text{mg}}(k) = P_{\text{mg}}(k) - P_{\text{mg}}(k-1)$.

Finally, the average power derivative (APD) is the average of the absolute value of the rate of change of the main grid power

$$\text{APD} = \frac{1}{T}\sum_{k=1}^{T} |\Delta P_{\text{mg}}(k)|. \tag{A7}$$

In this Appendix, the maximum and minimum values were taken over the whole simulation period.

References

1. Padhi, P.P.; Pati, R.K.; Nimje, A.A. Distributed generation: Impacts and cost analysis. *Int. J. Power Syst. Oper. Energy Manag.* **2012**, *1*, 91–96.
2. Hidalgo, R.; Abbey, C.; Joós, G. A review of active distribution networks enabling technologies. In Proceedings of the IEEE PES General Meeting, Providence, RI, USA, 25–29 July 2010; pp. 1–9. [CrossRef]

3. Abapour, S.; Zare, K.; Mohammadi-Ivatloo, B. Dynamic planning of distributed generation units in active distribution network. *IET Gener. Transm. Distrib.* **2015**, *9*, 1455–1463. [CrossRef]
4. Palizban, O.; Kauhaniemi, K.; Guerrero, J.M. Microgrids in active network management—Part II: System operation, power quality and protection. *Renew. Sustain. Energy Rev.* **2014**, *36*, 440–451. [CrossRef]
5. Mirzania, P.; Andrews, D.; Ford, A.; Maidment, G. Community Energy in the UK: The End or the Beginning of a Brighter Future? In Proceedings of the 1st International Conference on Energy Research and Social Science, Sitges, Spain, 5–7 April 2017.
6. Lefroy, J. Local Electricity Bill. Available online: https://powerforpeople.org.uk/wp-content/uploads/2019/03/Local-Electricity-Bill.pdf (accessed on 3 October 2019).
7. Fazeli, A.; Sumner, M.; Johnson, M.C.; Christopher, E. Real-time deterministic power flow control through dispatch of distributed energy resources. *IET Gener. Transm. Distrib.* **2015**, *9*, 2724–2735. [CrossRef]
8. Daud, A.K.; Ismail, M.S. Design of isolated hybrid systems minimizing costs and pollutant emissions. *Renew. Energy* **2012**, *44*, 215–224. [CrossRef]
9. Hosseinzadeh, M.; Salmasi, F.R. Power management of an isolated hybrid AC/DC microgrid with fuzzy control of battery banks. *IET Renew. Power Gener.* **2015**, *9*, 484–493. [CrossRef]
10. Arcos-Aviles, D.; Pascual, J.; Marroyo, L.; Sanchis, P.; Guinjoan, F. Fuzzy Logic-Based Energy Management System Design for Residential Grid-Connected Microgrids. *IEEE Trans. Smart Grid* **2018**, *9*, 530–543. [CrossRef]
11. Davies, R.; Sumner, M.; Christopher, E. Energy storage control for a small community microgrid. In Proceedings of the 7th IET International Conference on Power Electronics, Machines and Drives, Manchester, UK, 8–10 April 2014; pp. 1–6. [CrossRef]
12. Silvente, J.; Kopanos, G.M.; Pistikopoulos, E.N.; Espuña, A. A rolling horizon optimization framework for the simultaneous energy supply and demand planning in microgrids. *Appl. Energy* **2015**, *155*, 485–501. [CrossRef]
13. Palma-Behnke, R.; Benavides, C.; Lanas, F.; Severino, B.; Reyes, L.; Llanos, J.; Sáez, D. A Microgrid Energy Management System Based on the Rolling Horizon Strategy. *IEEE Trans. Smart Grid* **2013**, *4*, 996–1006. [CrossRef]
14. Parisio, A.; Rikos, E.; Glielmo, L. A Model Predictive Control Approach to Microgrid Operation Optimization. *IEEE Trans. Control Syst. Technol.* **2014**, *22*, 1813–1827. [CrossRef]
15. Olivares, D.E.; Cañizares, C.A.; Kazerani, M. A Centralized Energy Management System for Isolated Microgrids. *IEEE Trans. Smart Grid* **2014**, *5*, 1864–1875. [CrossRef]
16. Ji, Z.; Huang, X.; Xu, C.; Sun, H. Accelerated Model Predictive Control for Electric Vehicle Integrated Microgrid Energy Management: A Hybrid Robust and Stochastic Approach. *Energies* **2016**, *9*, 973. [CrossRef]
17. Li, Z.; Zang, C.; Zeng, P.; Yu, H. Combined Two-Stage Stochastic Programming and Receding Horizon Control Strategy for Microgrid Energy Management Considering Uncertainty. *Energies* **2016**, *9*, 499. [CrossRef]
18. Brandstetter, M.; Schirrer, A.; Miletić, M.; Henein, S.; Kozek, M.; Kupzog, F. Hierarchical Predictive Load Control in Smart Grids. *IEEE Trans. Smart Grid* **2017**, *8*, 190–199. [CrossRef]
19. Shayeghi, H.; Shahryari, E.; Moradzadeh, M.; Siano, P. A Survey on Microgrid Energy Management Considering Flexible Energy Sources. *Energies* **2019**, *12*, 2156. [CrossRef]
20. Lara, J.D.; Olivares, D.E.; Cañizares, C.A. Robust Energy Management of Isolated Microgrids. *IEEE Syst. J.* **2018**, 1–12. [CrossRef]
21. Wu, W.; Chen, J.; Zhang, B.; Sun, H. A Robust Wind Power Optimization Method for Look-Ahead Power Dispatch. *IEEE Trans. Sustain. Energy* **2014**, *5*, 507–515. [CrossRef]
22. Khodaei, A. Provisional Microgrids. *IEEE Trans. Smart Grid* **2015**, *6*, 1107–1115. [CrossRef]
23. Yi, J.; Lyons, P.F.; Davison, P.J.; Wang, P.; Taylor, P.C. Robust Scheduling Scheme for Energy Storage to Facilitate High Penetration of Renewables. *IEEE Trans. Sustain. Energy* **2016**, *7*, 797–807. [CrossRef]
24. Malysz, P.; Sirouspour, S.; Emadi, A. An Optimal Energy Storage Control Strategy for Grid-connected Microgrids. *IEEE Trans. Smart Grid* **2014**, *5*, 1785–1796. [CrossRef]
25. Valencia, F.; Collado, J.; Sáez, D.; Marín, L.G. Robust Energy Management System for a Microgrid Based on a Fuzzy Prediction Interval Model. *IEEE Trans. Smart Grid* **2016**, *7*, 1486–1494. [CrossRef]
26. Du, Y.; Pei, W.; Chen, N.; Ge, X.; Xiao, H. Real-time microgrid economic dispatch based on model predictive control strategy. *J. Mod. Power Syst. Clean Energy* **2017**, *5*, 787–796. [CrossRef]

27. Zhang, Y.; Fu, L.; Zhu, W.; Bao, X.; Liu, C. Robust model predictive control for optimal energy management of island microgrids with uncertainties. *Energy* **2018**, *164*, 1229–1241. [CrossRef]
28. Velarde, P.; Maestre, J.M.; Ocampo-Martinez, C.; Bordons, C. Application of robust model predictive control to a renewable hydrogen-based microgrid. In Proceedings of the 2016 European Control Conference (ECC), Aalborg, Denmark, 29 June–1 July 2016; pp. 1209–1214. [CrossRef]
29. Pereira, M.; de la Peña, D.M.; Limon, D. Robust economic model predictive control of a community micro-grid. *Renew. Energy* **2017**, *100*, 3–17. [CrossRef]
30. Fazeli, A.; Sumner, M.; Christopher, E.; Johnson, M. Power flow control for power and voltage management in future smart energy communities. In Proceedings of the 3rd Renewable Power Generation Conference, Naples, Italy, 24–25 September 2014; pp. 1–6. [CrossRef]
31. Elkazaz, M.; Sumner, M.; Thomas, D. Real-Time Energy Management for a Small Scale PV-Battery Microgrid: Modeling, Design, and Experimental Verification. *Energies* **2019**, *12*, 2712. [CrossRef]
32. Plett, G.L. High-performance battery-pack power estimation using a dynamic cell model. *IEEE Trans. Veh. Technol.* **2004**, *53*, 1586–1593. [CrossRef]
33. Sun, F.; Xiong, R.; He, H.; Li, W.; Aussems, J.E. Model-based dynamic multi-parameter method for peak power estimation of lithium–ion batteries. *Appl. Energy* **2012**, *96*, 378–386. [CrossRef]
34. Pérez, A.; Moreno, R.; Moreira, R.; Orchard, M.; Strbac, G. Effect of Battery Degradation on Multi-Service Portfolios of Energy Storage. *IEEE Trans. Sustain. Energy* **2016**, *7*, 1718–1729. [CrossRef]
35. Der Merwe, R.V.; Wan, E.A. The square-root unscented Kalman filter for state and parameter-estimation. In Proceedings of the 2001 IEEE International Conference on Acoustics, Speech, and Signal Processing, Salt Lake City, UT, USA, 7–11 May 2001; Volume 6, pp. 3461–3464. [CrossRef]
36. Tampier, C.; Pérez, A.; Jaramillo, F.; Quintero, V.; Orchard, M.; Silva, J. Lithium-ion battery end-of-discharge time estimation and prognosis based on Bayesian algorithms and outer feedback correction loops: A comparative analysis. In Proceedings of the Annual Conference of the Prognostics and Health Management Society, San Diego, CA, USA, 18–24 October 2015; Volume 6, pp. 182–195.
37. Burgos, C.; Sáez, D.; Orchard, M.E.; Cárdenas, R. Fuzzy modelling for the state-of-charge estimation of lead-acid batteries. *J. Power Sources* **2015**, *274*, 355–366. [CrossRef]
38. Pola, D.A.; Navarrete, H.F.; Orchard, M.E.; Rabié, R.S.; Cerda, M.A.; Olivares, B.E.; Silva, J.F.; Espinoza, P.A.; Pérez, A. Particle-Filtering-Based Discharge Time Prognosis for Lithium-Ion Batteries with a Statistical Characterization of Use Profiles. *IEEE Trans. Reliab.* **2015**, *64*, 710–720. [CrossRef]
39. Sáez, D.; Ávila, F.; Olivares, D.; Cañizares, C.; Marín, L. Fuzzy Prediction Interval Models for Forecasting Renewable Resources and Loads in Microgrids. *IEEE Trans. Smart Grid* **2015**, *6*, 548–556. [CrossRef]
40. Sáez, D.; Zuniga, R. Cluster optimization for Takagi & Sugeno fuzzy models and its application to a combined cycle power plant boiler. In Proceedings of the 2004 American Control Conference, Boston, MA, USA, 30 June–2 July 2004; Volume 2, pp. 1776–1781. [CrossRef]
41. Babuška, R. *Fuzzy Modeling for Control*, 1st ed.; Kluwer Academic Publishers: Boston, MA, USA, 1998; p. 288. [CrossRef]
42. Kouvaritakis, B.; Cannon, M. *Model Predictive Control: Classical, Robust and Stochastic*, 1st ed.; Springer: Berlin, Germany, 2016; p. 384. [CrossRef]
43. Goulart, P.J.; Kerrigan, E.C.; Maciejowski, J.M. Optimization over state feedback policies for robust control with constraints. *Automatica* **2006**, *42*, 523–533. [CrossRef]
44. Kouvaritakis, B.; Cannon, M.; Muñoz-Carpintero, D. Efficient prediction strategies for disturbance compensation in stochastic MPC. *Int. J. Syst. Sci.* **2013**, *44*, 1344–1353. [CrossRef]
45. Richardson, I.; Thomson, M. One-Minute Resolution Domestic Electricity Use Data, 2008–2009. *Colch. Essex UK Data Arch.* **2010**. [CrossRef]
46. Green Energy UK. *Tide—A New Way to Take Control*; Green Energy UK: Ware, UK. Available online: https://www.greenenergyuk.com/Tide (accessed on 3 October 2019).
47. Pascual, J.; Barricarte, J.; Sanchis, P.; Marroyo, L. Energy management strategy for a renewable-based residential microgrid with generation and demand forecasting. *Appl. Energy* **2015**, *158*, 12–25. [CrossRef]
48. Parra, D.; Norman, S.A.; Walker, G.S.; Gillott, M. Optimum community energy storage for renewable energy and demand load management. *Appl. Energy* **2017**, *200*, 358–369. [CrossRef]

49. Zahboune, H.; Zouggar, S.; Krajacic, G.; Varbanov, P.S.; Elhafyani, M.; Ziani, E. Optimal hybrid renewable energy design in autonomous system using Modified Electric System Cascade Analysis and Homer software. *Energy Convers. Manag.* **2016**, *126*, 909–922. [CrossRef]
50. Bilal, B.O.; Sambou, V.; Ndiaye, P.A.; Kébé, C.M.F.; Ndongo, M. Multi-objective design of PV-wind-batteries hybrid systems by minimizing the annualized cost system and the loss of power supply probability (LPSP). In Proceedings of the 2013 IEEE International Conference on Industrial Technology (ICIT), Cape Town, South Africa, 25–28 Feburary 2013; pp. 861–868. [CrossRef]

Article

An Evaluation of the Economic and Resilience Benefits of a Microgrid in Northampton, Massachusetts

Patrick Balducci [1,*], Kendall Mongird [1], Di Wu [2], Dexin Wang [2], Vanshika Fotedar [1] and Robert Dahowski [2]

1 Pacific Northwest National Laboratory (PNNL), Portland, OR 97204, USA; kendall.mongird@pnnl.gov (K.M.); vanshika.fotedar@pnnl.gov (V.F.)

2 Pacific Northwest National Laboratory (PNNL), Richland, WA 99352, USA; Di.Wu@pnnl.gov (D.W.); dexin.wang@pnnl.gov (D.W.); bob.dahowski@pnnl.gov (R.D.)

* Correspondence: Patrick.balducci@pnnl.gov; Tel.: +1-(503)-679-7316

Received: 30 July 2020; Accepted: 9 September 2020; Published: 14 September 2020

Abstract: Recent developments and advances in distributed energy resource (DER) technologies make them valuable assets in microgrids. This paper presents an innovative evaluation framework for microgrid assets to capture economic benefits from various grid and behind-the-meter services in grid-connecting mode and resilience benefits in islanding mode. In particular, a linear programming formulation is used to model different services and DER operational constraints to determine the optimal DER dispatch to maximize economic benefits. For the resiliency analysis, a stochastic evaluation procedure is proposed to explicitly quantify the microgrid survivability against a random outage, considering uncertainties associated with photovoltaic (PV) generation, system load, and distributed generator failures. Optimal coordination strategies are developed to minimize unserved energy and improve system survivability, considering different levels of system connectedness. The proposed framework has been applied to evaluate a proposed microgrid in Northampton, Massachusetts that would link the Northampton Department of Public Works, Cooley Dickenson Hospital, and Smith Vocational Area High School. The findings of this analysis indicate that over a 20-year economic life, a 441 kW/441 kWh battery energy storage system, and 386 kW PV solar array can generate $2.5 million in present value benefits, yielding a 1.16 return on investment ratio. Results of this study also show that forming a microgrid generally improves system survivability, but the resilience performance of individual facilities varies depending on power-sharing strategies.

Keywords: economic analysis; energy storage systems; microgrid; resilience

---

1. Introduction

Battery energy storage systems (BESSs) operating within microgrids and in isolation can improve the electrical grid's operational flexibility [1]. The ability of microgrids to act as an isolated grid capable of islanding away from a region's interconnected electrical grid can also enhance resilience for parties covered by microgrid assets [2]. In recent years, an extensive body of literature has formed around questions addressing the economic and resilience benefits of microgrids.

Optimal sizing and scheduling of remote microgrid assets operating in isolation have been addressed in several recent studies. For example, [3] proposes a risk-based stochastic approach to optimally scheduling microgrid assets. In [4], a Monte Carlo simulation-based stochastic programming approach is used to define an optimal design for a remote microgrid based on reliability and cost criteria. Based on the assessment of a subset of representative days, annualized costs and fuel consumption rates are minimized through the optimal scaling and operation of microgrid assets in [5]. The coordination of wind turbines, an energy storage system and a diesel generator operating in a remote microgrid is achieved through a dynamic programming method in [6].

There have also been several articles recently published that focus on maximizing the economic benefits of distributed energy resources (DERs) operating behind-the-meter (BTM) under normal conditions while in grid-connected mode. A mixed integer programming method is proposed in [7] for optimally scaling battery energy storage systems (BESS) to minimize total investment and minimize the deviation between scheduled and actual imported power to the end user. Optimal DER planning with the objective of minimizing fuel consumption while complying with state renewable energy mandates at industrial sites and campus communities is addressed in [8]. The authors address this problem through the use of Lagrange Multipliers with consideration given to numerous factors, including climate conditions, environmental regulations and energy resource availability. In [9], the economic returns of a BESS are evaluated while operating BTM at a military base in the Independent System Operator New England (ISO-NE) territory and when being used to minimize energy and demand charges, reduce capacity payments and participate in regional demand response programs. This study evaluates a range of BESS power and energy capacities in order to optimally scale the system based on the landscape of economic opportunity.

There is a smaller subset of the literature base dedicated to simultaneously evaluating the economic and resiliency performance of microgrids. In [10], the authors propose a resiliency-oriented microgrid optimal scheduling method. Under normal operation, it minimizes the operating costs of the microgrid. For resiliency analyses during outages, it solves a max–min problem to minimize the worst-case energy imbalance. Although both economic and resiliency are covered in the analyses, limited bill reduction benefits (only time-of-use pricing) is considered and no asset failure is modeled. The authors of [11] evaluate both the economic and resiliency benefits of DERs, including BESS and photovoltaic (PV) units, integrated into large buildings. This study evaluates the energy cost reductions associated with microgrid operations but does not include the value of outage mitigation. In [12], an approach is proposed for optimally scaling a solar-plus-storage system in order to minimize financial losses caused by disruptions resulting from grid outages. In [13], a method is proposed for incorporating the value of resilience delivered by BESSs and PV systems into investment decisions for building-scale microgrids.

The main contributions of this paper are twofold.

- Unlike many existing microgrid planning and expansion studies that consider either economic or resilience benefits, this paper presents an innovative evaluation framework for microgrid assets to capture economic benefits from various grid and BTM services in grid-connecting mode and resilience benefits in islanding mode.
- As for the resiliency analysis, an innovative method was proposed to explicitly quantify the microgrid survivability against a random outage. The method proposed here models the stochastic nature of PV generation, system load, and potential distributed generator (DG) failure during an outage. In addition, to optimally use the microgrid assets, coordination strategies are developed to minimize unserved energy and improve system survivability, considering different levels of system connectedness. In particular, rules are proposed to fairly distribute unserved energy among facilities when load shedding becomes inevitable.

The proposed framework has been applied to evaluate a proposed microgrid in Northampton, Massachusetts that would link the Northampton Department of Public Works (DPW), Cooley Dickenson Hospital (CDH), and Smith Vocational Area High School (SVAHS). The findings of this analysis indicate that over a 20-year economic life, a 441 kW/441 kWh BESS and 386 kW PV solar array can generate $2.5 million in present value benefits, yielding a 1.16 return on investment (ROI) ratio. Results of this study also show that forming a microgrid generally improves system survivability, but the resilience performance of individual facilities varies depending on power-sharing strategies.

The rest of this paper is organized as follows. In Section 2, we present an overview of the Northampton Microgrid Project (NMP) and outline the methods and data used to perform the economic and resilience evaluations of the NMP. Using the proposed methods, NMP economic returns were defined and resiliency analysis was performed for a number of representative scenarios characterized by season, system connectedness, DER availability, and outage durations. Analysis results are presented in Section 3. Section 4 presents a discussion of the results within the context of other recent microgrid assessments. Section 5 presents study conclusions.

2. Materials and Methods

The City of Northampton, Massachusetts, in partnership with the CDH, was awarded a $3,078,960 grant from the Massachusetts Department of Energy Resources (DOER) to develop a microgrid as part of its Community Clean Energy Resiliency Initiative. The grant stemmed from a DOER-supported analysis of three linked Northampton facilities. These facilities include: CDH, the SVAHS, and the Northampton DPW. SVAHS is a 10-building school campus that serves approximately 600 people a day. The facility also acts as the regional Red Cross Emergency Shelter while providing emergency overflow services for CDH. The Northampton DPW provides support for a number of critical city functions, including emergency services radio communication, flood control, stormwater systems, and clean water processing and delivery. The CDH campus includes emergency services facilities, as well as a number of other patient care facilities and serves approximately 2500 people each day under normal conditions and 1000 during an emergency. All three facilities are shown in Figure 1.

Figure 1. Satellite Google Earth image of CDH, SVAHS, and Northampton DPW. Map data: Apache License, Version 2.0, Google, 2020 [14,15].

The NMP, which is in the design phase, is planned to be comprised of PV, energy storage, a biomass facility, and diesel generators. During normal operations these assets will be operated in isolation in order to minimize the electricity bills of each separate facility. In the event of an outage these assets will be aggregated to create an islanded microgrid capable of enhancing the resiliency for all three facilities.

Currently, CDH operates a 275 kW biomass steam turbine generation unit and a 175 kW turbine supporting a 700 ton absorption chiller. CDH is considering the addition of a 386 kW PV array, as well as a 441 kW/441 kWh BESS. Under current NMP plans, the BESS included in the microgrid will use a lithium-ion battery technology. This was chosen over a sealed lead acid battery due to the sensitivity of the latter to deep discharges, which the battery would face during outage events. While a lithium-ion battery's economic life would not typically be expected to exceed 10 years, this analysis assumes that an augmentation plan, or extended warranty, is purchased to extend its life to 20 years. Additionally,

SVAHS currently has a 106 kW PV array installed and existing diesel generators at the three sites include:

1. 155 kW at SVAHS;
2. 40 kW at the DPW; and,
3. 2.4 MW at the CDH (3 units at 800 kW each).

Figure 2 presents the aforementioned power components as currently configured at all three facilities when operating under either normal or emergency conditions. The enhanced microgrid as it would operate during power outages is presented in the far-right box. Note that the biomass facilities were excluded from the resiliency analysis because they would trip off during voltage sags and would therefore be unsuitable for providing power when operated in islanded mode.

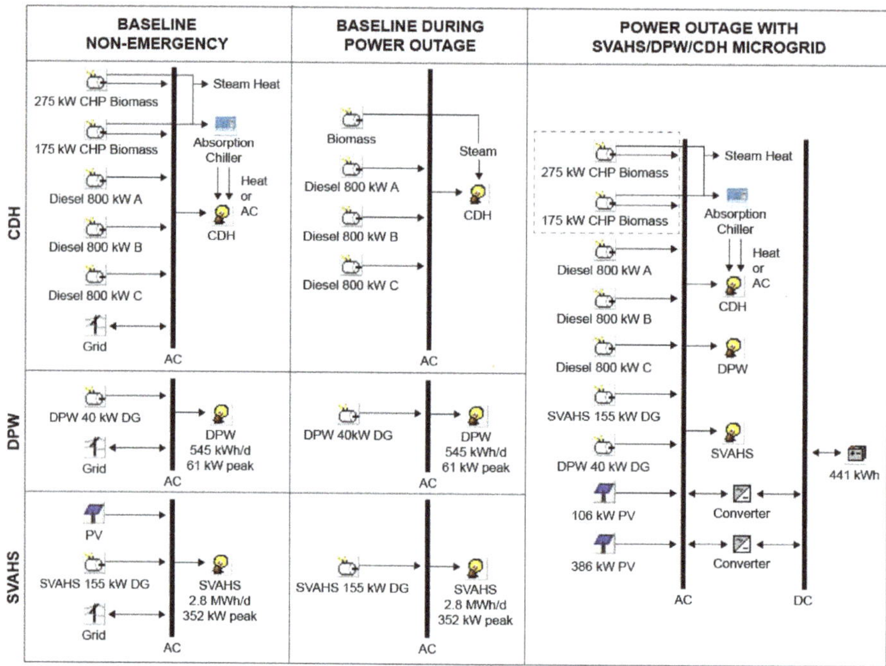

Figure 2. Existing and planned power components of the Northampton microgrid project.

The objectives in conducting this research have been to define and model the financial and resilience opportunities available to the City of Northampton, Massachusetts associated with installing a microgrid connecting the Northampton DPW, CDH, and SVAHS. This section presents the methodologies employed on our economic and resiliency assessments.

2.1. Economic Methodology

The energy assets proposed for the microgrid were modeled to enable simulation of Northampton microgrid operations. Bundling of services (providing multiple services over a set period) was analyzed and the improvement in the overall economics of the microgrid evaluated.

The Battery Storage Evaluation Tool (BSET—PNNL, Richland, WA, USA), which was developed by this project team for Pacific Northwest National Laboratory (PNNL), was used to estimate the economic benefits of a BESS and a PV unit. A large linear programming problem was formulated to maximize the total benefits from multiple value streams, considering various system- and component-level constraints, such as power balancing, BESS charging/discharging limits, state of charge dynamics, models for each individual use cases and their couplings. The mathematical basis of the BSET model was presented in [16].

Economic benefits associated with NMP operations were modeled in BSET for a one-year period. BSET was used to determine the value of each use case or service and the number of hours the BESS would optimally engage in providing each service. After microgrid operations were modeled for one year, the results were used for several purposes, including: (a) evaluating the economic benefits of the planned BESS and PV additions, and (b) assessing microgrid capacity to improve resiliency.

This economic assessment evaluates each of the services outlined below:

- National Grid demand charge reduction
- National Grid demand response program participation
- ISO-NE installed capacity tag reduction
- Outage mitigation
- Energy purchase reduction through PV array production
- PV renewable energy credits (RECs)

Each of the services above and their associated methodologies will be described later in this section. Prior to this discussion, we focus first on the load analysis that supported the economic optimization.

2.1.1. Load Analysis

The CDH and SVAHS each have historical interval electricity data available from the utility to assist with understanding electric loads, variation, and sizing of the microgrid system (Figures 3 and 4). However, the Northampton DPW facility did not have a record of electrical load at such intervals, and only monthly billing totals were available. Therefore, the Facility Energy Decision System (FEDS—PNNL, Richland, WA, USA) software was used to derive representative load profiles for the DPW buildings.

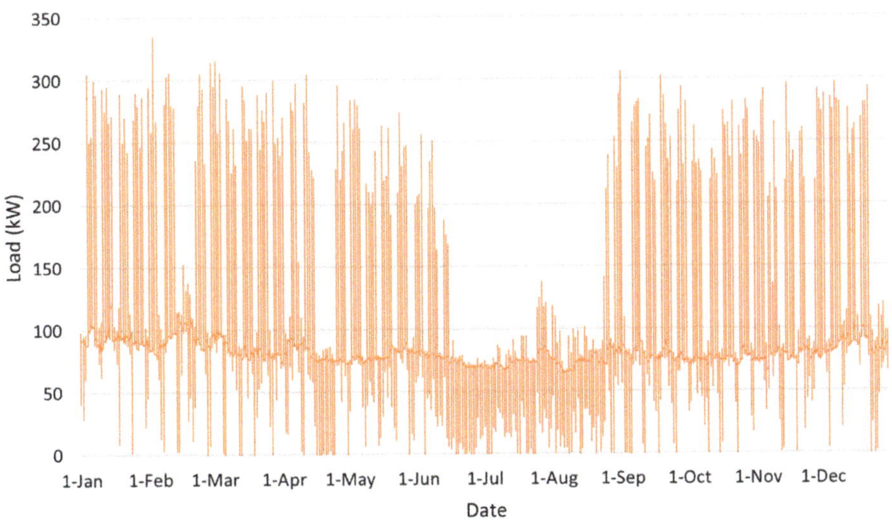

Figure 3. SVAHS hourly peak load, 2016.

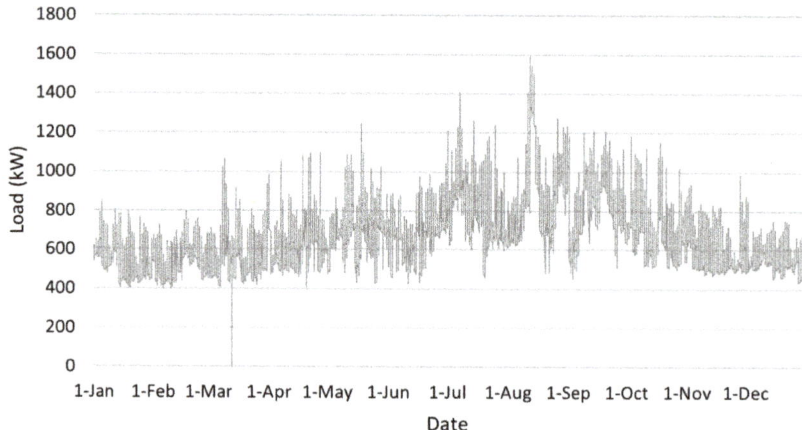

Figure 4. CDH hourly peak load, 2016.

FEDS is a building energy simulation and analysis tool developed by PNNL, which has been used extensively to understand energy use and opportunities for cost-effective energy saving projects within single building and multi-building campuses across the U.S. and around the world. FEDS relies on detailed building system inputs and is backed by a sophisticated inference engine to allow for relatively quick yet robust modeling of building systems, loads, and energy use. Key inputs range from building types, size, vintage, location, hourly weather data, plus details on occupancy, building envelope, lighting, heating, cooling, ventilation (HVAC), and hot water systems. Based on the inputs and inferred parameters, FEDS performs an hourly simulation of energy use at each technology, end use, and building level. An hourly profile for energy use for each fuel consumed is one of the outputs and is used to represent the electricity load profile for buildings for resilience assessments and input for sizing of backup power or microgrid systems [17].

There are two buildings at the Northampton DPW site: a small administration building and a garage. These are shown in the satellite image in Figure 5. The administration building is 3868 ft^2 and was built in 1973. The 20,700 ft^2 garage is an old trolley barn built around the start of the 20th century, and provides vehicle bays, shops, and some limited office space. Without the ability to perform an on-site walk-through, available data on occupancy, HVAC, and hot water systems were provided by City of Northampton staff, along with the result of recent assessments that included information on envelope characteristics, lighting types, and counts. Google Earth satellite and street view imagery (Apache Software Foundation, Wakefield, MA, USA) was used to help validate select data, and fill in gaps regarding modeling geometries, construction characteristics, window fraction, and more. Energy billing data was also provided by the city, including monthly electricity and natural gas consumption for recent years.

FEDS models were developed for each of the two DPW buildings based on data provided and gathered from available sources. One of the unique features of FEDS is that it requires only a relatively small set of input parameters in order to begin modeling. Parameters not specified are automatically inferred by FEDS to the most likely value based on all of the specified inputs (including but not limited to building type, size, vintage, location, use, and technology types). These inferences are based on a combination of building survey data, building codes, equipment standards, along with decades of experience with buildings and building systems. All inferred values can be reviewed, updated or overridden by the modeler.

Screen captures of the two DPW FEDS models are shown in Figure 6. They show high-level building inputs for each model, along with some of the details modeled for lighting, HVAC, and hot water systems. Additionally, FEDS is able to infer base plug and process loads for each building based on use type. For the DPW garage, two distinct use areas were modeled covering the garage/shop and office areas.

Once developed, the FEDS models were reviewed and calibrated to recent historical meter data to ensure a representative simulation and resulting load profiles. A weather adjustment was applied to the typical meteorological year weather data used by the FEDS simulation, based on the actual weather trends for each calibration year. The models were calibrated to both 2017 and 2018 metered consumption data for each building. Both models were calibrated to match the annual energy consumption while maintaining similar monthly magnitude and trends. Results of the calibration comparing monthly electricity consumption from the FEDS simulation compared to actual consumption are highlighted by Figure 7.

Once the building models were satisfactorily calibrated, each building model was simulated with FEDS to produce a representative hourly load profile covering an entire year. A typical meteorological weather year was applied to represent typical historical weather patterns against the calibrated models. The resulting annual hourly electric load profile for the two DPW buildings, as simulated by FEDS, is shown in Figure 8.

Figure 5. Satellite image from Google Earth shows location of Northampton DPW buildings (garage on top and administrative building on bottom). Map data: Apache License, Version 2.0, Google, 2020 [14,18].

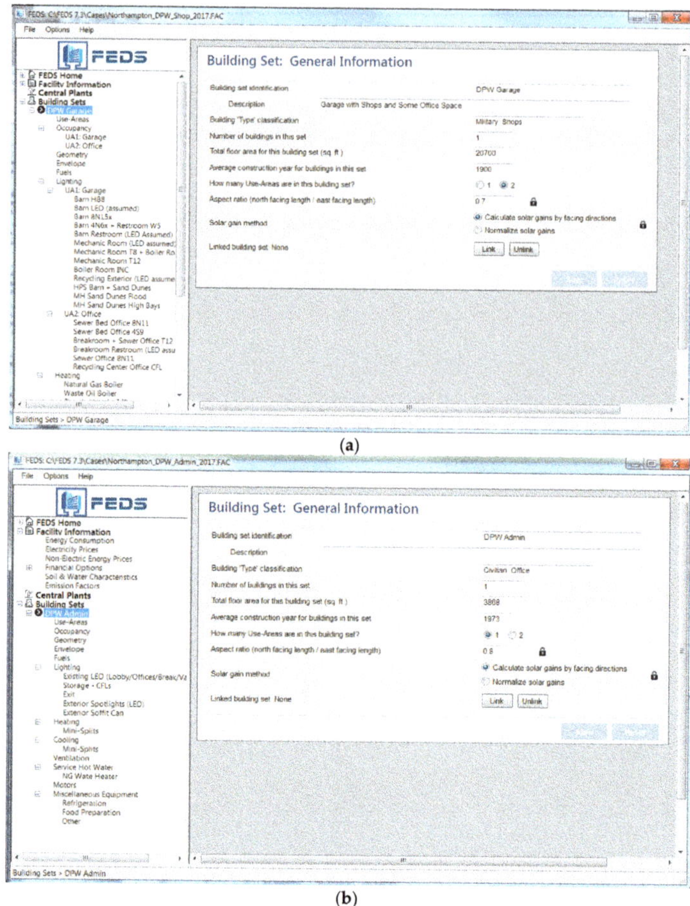

Figure 6. FEDS screen shots show the DPW garage (**a**) and administrative building (**b**) models. Created via the FEDS software [17].

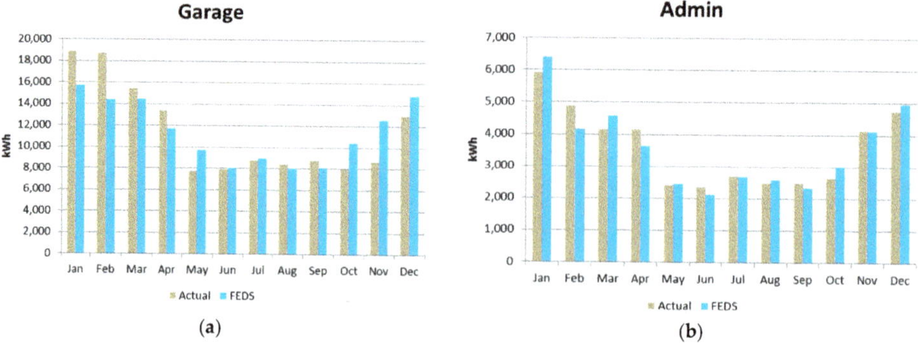

Figure 7. Results of FEDS simulated model calibration show monthly electricity consumption vs. actuals for the DPW garage (**a**) and administrative building (**b**).

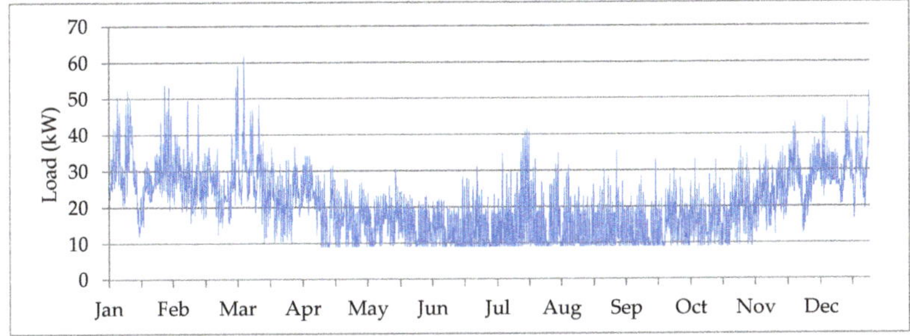

Figure 8. Combined electric load profile for DPW.

2.1.2. Use Cases

This section defines the use cases that could result in financial benefits to the NMP partners. The methods used to assign value to the benefits tied to both BESS and PV operations are defined.

National Grid Demand Charge Reduction

The three sites that comprise the microgrid all purchase energy from National Grid—a large distribution utility that serves customers in Massachusetts and beyond. Every month, each of these microgrid members faces a demand charge of $5.76/kW on their energy bills, which is correlated to their single highest 15 min load between the hours of 8:00 am and 9:00 pm, Monday through Friday. If the BESS reduced the peak load in a given month by 200 kW by discharging at the appropriate times, then $1152 is obtained in value each month. Of the three sites of the microgrid project, a CDH account had the largest peaks in summer months and, therefore, stands to gain the most benefit through demand charge reduction. Therefore, the battery was modeled at CDH using BSET and historic load data defined in Section 2.1.1.

National Grid Demand Response Program Participation

A BTM BESS could participate in the Connected Solutions Program, which compensates commercial and industrial customers to curtail their energy when the ISO-NE system is forecasted to be at its peak. A participating BESS would be compensated for the amount of energy curtailed on a pay-for-performance basis.

The program offers three options to participate:

1. Targeted dispatch to reduce load at the peak hour of the year (two to eight dispatch events per summer), valued at $35/kW—summer.
2. Daily dispatch to reduce load at the peak hour of the year and during daily peaks in July and August (30–60 dispatch events per summer), valued at $200/kW—summer.
3. Winter dispatch to reduce load during five peak hours of the winter, valued at $25/kW—winter.

The events last between two to three hours for daily dispatch and three hours for targeted dispatch. An event can happen anytime between 2:00–7:00 pm on non-holiday weekdays during summer or winter. For a 441 kW/441 kWh BESS, only one-third of its capacity could be bid into the demand response program because the basis of compensation is the average energy discharged during the three-hour window.

Based on input directly from a distribution utility in Massachusetts, we assume the daily dispatch events will start on 1 July and would be held on every non-holiday weekday from 4:00–7:00 pm until all 60 calls take place. The summer and winter targeted dispatch are typically scheduled to occur on the days when the utility expects the summer and winter peaks to happen, respectively. For the winter program, we assume that calls occur on the five peak days registered during winter, which all fall between 1 January and 7 January.

We assume that the BESS would participate in the daily dispatch and targeted winter dispatch program. While additional benefits could be obtained by participating in the summer targeted dispatch program, the BESS would have to provide twice the capacity to obtain those benefits simultaneously while also participating in the daily dispatch program. Therefore, we exclude the summer targeted dispatch program.

ISO-NE Installed Capacity Tag Reduction

The members of the Northampton microgrid face a monthly charge called the installed capacity (ICAP) tag. The charge amount is dependent on their measured load during ISO-NE's peak hour each year. This calculation is made only once a year but affects the following year's monthly bills. That is, an ICAP calculated in July of 2020 will not take effect until June 2021 at the start of the next billing season.

The price for the ICAP follows an annual schedule, multiple years of which are shown in Table 1. Forecasts beyond the actual values provided in the table were provided by National Grid for an analysis of a BESS deployed on Nantucket Island, Massachusetts. More information regarding this forecast can be found in [19].

Table 1. Independent System Operator New England (ISO-NE) installed capacity (ICAP) tag rates 2018–2023.

Year Start	Year End	Capacity ($/kW—Month)
June-2018	May-2019	9.55
June-2019	May-2020	7.03
June-2020	May-2021	5.30
June-2021	May-2022	4.63
June-2022	May-2023	3.80
June-2023	May-2024	2.00

Typically, the ISO-NE annual peak occurs in June, July, or August and usually between the hours of 2:00–5:00 pm. It is also important to note that it will almost certainly be coincident with one of the National Grid demand response events described previously. That is, by hitting all National Grid demand response events during a summer, it is almost guaranteed to hit the ICAP tag date/time, picking up double benefits for the same load reduction. Table 2 shows historic values of when the ICAP tag has been called.

Table 2. ISO-NE historic ICAP tag dates.

Year	Date	Time
2001	9 August	2 pm
2002	14 August	2 pm
2003	22 August	2 pm
2004	30 August	3 pm
2005	27 July	2 pm
2006	2 August	2 pm
2007	3 August	2 pm
2008	10 June	2 pm
2009	18 August	2 pm
2010	6 July	2 pm
2011	22 July	2 pm
2012	17 July	4 pm
2013	19 July	4 pm
2014	2 July	2 pm
2015	29 July	4 pm
2016	12 August	2 pm
2017	13 June	4 pm
2018	29 August	4 pm
2019	30 July	5 pm

Reductions in load tied to PV production are also included in the value obtained for this use case. Solar production reduces the registered demand during the system-wide peak, thereby reducing the total kW-demand that forms the basis of the ICAP tag.

Outage Mitigation

In the event of an outage, the BESS has the capability to effectively operate in an islanded mode. This operation would be monetized in terms of the value of lost load (VOLL). To estimate the benefits that can be derived from outage mitigation, historical events were examined at DPW and SVAHS. CDH does not experience outages due to the presence of onsite generation, but outages do occur at the Northampton DPW and SVAHS. From these historical outage occurrences, the timing and duration of the outages were defined for these facilities.

In order to assign monetary values to reducing or eliminating potential outages, the findings of [20] from Lawrence Berkeley National Laboratory were used. This process estimates costs based on customer group (residential, small commercial and industrial (<50,000 annual kWh load) and medium/large commercial and industrial) and the duration of the outage. Figure 9 presents an example of the trendline used to estimate the cost for different lengths of outage for medium/large commercial and industrial customers.

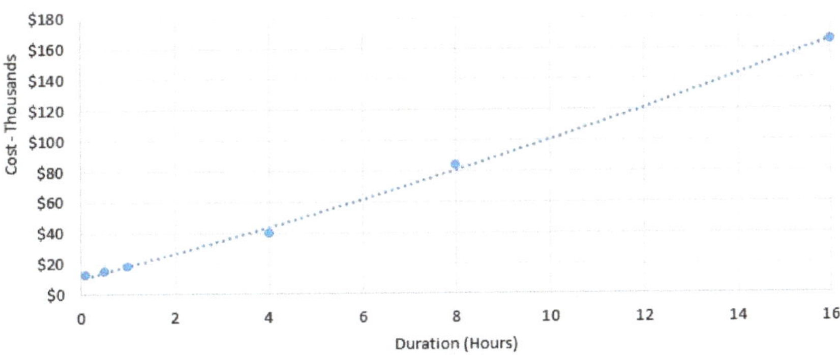

Figure 9. Value of lost load by outage duration to medium/large commercial and industrial customers.

Outages were modeled with no foresight, meaning that the BESS would not be prepared for outages but rather would use available energy based on its state of charge, along with energy output from the PV unit, to address any outage. The savings to customers served by the microgrid is estimated to be $13,085 annually based on the aforementioned cost assumptions and the duration and frequency of outages.

Energy Purchase Reductions from PV Production

As previously described, the members of the Northampton microgrid purchase their energy from National Grid at time of use rates. The installation of the 386 kW solar array at CDH will reduce their monthly electricity purchases for all hours when the array generates energy.

To evaluate the potential energy production from solar PV panels in Northampton, a number of solar profiles were used with an established solar PV production model. The solar data was gathered from the National Solar Radiation Database (NSRDB)—specifically, the Physical Solar Model (PSM) dataset. This dataset provides both a large temporal coverage over 20 years of data, as well as relatively fine spatial resolution (4 × 4 km) for all locations in the continental U.S. (Figure 10). The data used in this evaluation is hourly (though 30-min data is also available) [21].

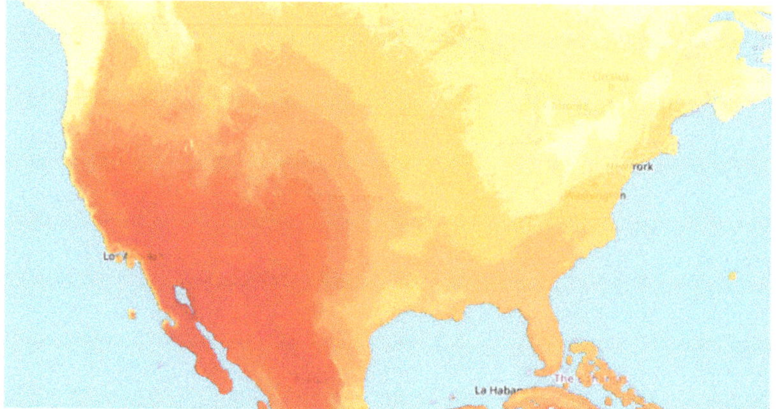

Figure 10. NSRDB data viewer. Created via the NSRDB [21].

The data itself is not measured data but rather modeled data based on satellite observations and models that estimate the impact on solar parameters due to atmospheric conditions. A comprehensive evaluation of the modeling techniques can be found at [22], which shows that modeled PSM data has a mean bias error of ±5% to ±10%, depending on the solar parameter. Though higher accuracy is always more desirable, for the purposes of this analysis this data was deemed sufficient.

Based on the PV modeling, the solar is expected to produce approximately 406,228 kWh of energy each year. An annual rate of degradation of 0.5% was assumed.

PV RECs

RECs are renewable energy certificates that solar owners receive at a rate of 1 REC per MWh of produced solar energy. These certificates are oftentimes sold to utilities so that they may meet renewable portfolio standards outlined in their state. Electricity providers must obtain RECs as proof that they have met guidelines for renewable generation. Various programs exist across various states that offer higher compensation for RECs, Massachusetts included. However, the 386 kW solar array would currently only be able to qualify for Class I RECs, which could be sold in the Class I REC auction.

Figure 11 shows recent Class I REC prices, including those in Massachusetts, from Power Advisory, LLC. Using these auction prices in combination with the solar production analysis, value can be estimated [23].

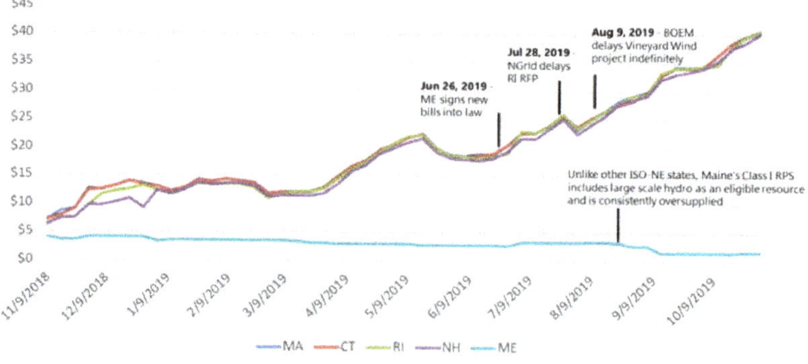

Figure 11. ISO-NE class 1 REC prices ($/REC).

2.1.3. Valuation Model

PNNL used BSET to model the operation of a BESS and a PV unit on an hourly basis over a year to evaluate their benefits. For each service, revenue or avoided costs were defined on an hourly basis using the methods outlined in the previous section, along with any power reservation or energy requirements to satisfy the demands of each use case. BSET was then used to co-optimize the benefits among these services, subject to the technical constraints of the BESS. Model output includes the value of, and the number of hours the BESS would be engaged in providing, each service.

2.1.4. Costs and Financial Parameters

The individual cost components of the project are broken down in Tables 3 and 4. Table 3 presents upfront costs incurred at the outset of the project except for fixed operations and maintenance (O & M) costs, which are incurred on an annual basis. For Table 4, it is assumed that major maintenance is conducted in Years 7 and 14 with full battery module replacement in Year 10. Maintenance and replacement are required to ensure battery operation for 20 years—lithium-ion systems typically have a 10-year usable life otherwise.

Table 3. Initial costs of energy storage systems.

Item	Cost
DC Modules and Battery Management System	$226,013
Power Conversion System	$187,425
Power Control System	$44,100
Electrical Balance of Plant	$44,100
Construction and Commissioning	$70,560
Fixed O & M Cost (per year)	$3969
Total	$576,167

Table 4. Major maintenance and battery replacement costs for lithium-ion systems.

Item	Cost
Major Maintenance ($/kW)	$275
Battery Replacement ($/kWh)	$100

The cost of the PV array was estimated based on the median $2.95/watt cost from commercial solar producers in Massachusetts, estimated by the Massachusetts Clean Energy Center for 2018–2019 [24]. Based on this cost estimate, the total cost of the 386 kW array is estimated to be $1.14 million. A detailed pro forma for the BESS and PV systems were prepared to estimate full costs. The financial parameters used in the pro forma are presented in Table 5.

Table 5. Financial parameters used in pro forma.

Parameter	Value
Energy Storage Book Life	20 years
Inflation Rate	2.5%
Benefit Growth Rate	2.64%
Discount Rate	2.2%
Insurance Rate	0.7%
PV Degradation Rate	0.5%/year

Based on the combination of costs and financial parameters outlined previously in this section, the researchers were able to produce a pro forma that accounted for full system costs. For the BESS, total costs for Northampton amount to $1,024,299. For the solar PV, the present value cost is $1,298,844 based on the median $/watt rate for commercial PV installations in Massachusetts from 2018–2019 [24].

2.2. Resilience Analysis Methodology

A resiliency analysis was conducted in addition to the economic analysis. The objective of the resiliency analysis is to understand the system survivability against a random outage with different starting times and durations in summer and winter seasons, considering different system connectedness, DER availability, and impacts of DG failure rates. The benefits of microgrid operations from the perspective of resiliency is evaluated by comparing the survivability across various scenarios. This section presents the key assumptions and modeling method.

2.2.1. Resiliency Modeling Methodology

To evaluate system resiliency, we randomly generated a large number of outages for each scenario characterized by changes in outage duration, season, system configuration, and DER availability. For each outage event, we formulated an optimal dispatch problem to minimize load shedding. The outputs of the optimal dispatch problem include whether there is any load shedding, DER operating levels, and hours with load shedding for each facility. The results from all outage events are used to calculate the survivability (the probability of the system to survive a random outage) and statistics of unserved energy at the facility level.

In the optimal dispatch formulation, we model the physical capability and operation of each DER:

- For the BESS, we adopt the linear model developed in [16] to capture power and energy limits, charging/discharging efficiencies, and dynamics of battery energy state.
- For PV, the same normalized power output in economic analysis is scaled by the installed capacity to calculate its power output in maximum power point tracking mode. Please note that dump energy from PV is allowed when the system cannot absorb all power from PV.
- For DG, a simplified linear model is used to represent fuel efficiency and power output limits. DG failure rates under different operating conditions are also considered.

Variables of hourly unserved energy are introduced to capture hourly load shedding in each outage event. A power balancing constraint is introduced for each hour at either the system or facility level, where the total supply from DERs plus unserved energy is equal to the load. For each outage event, the total fuel consumption cannot exceed the onsite fuel storage capacity. The objective of the optimal dispatch problem is, therefore, to minimize the total unserved energy during an outage. When there are multiple optimal solutions, the one with the least fuel consumption is preferred. Additional modeling details are provided as follows.

2.2.2. System Connectedness

We consider three levels of connectedness in the resiliency analysis:

1. No microgrid: This is the case where power-sharing is not enabled among the three facilities. During an outage, each facility purely relies on its local resources to support its load until power balance is maintained at the facility level.
2. Limited microgrid: In this case, power-sharing is allowed but the three DGs at CDH are reserved for the local load. Only the power from PV and the BESS can be exported to support the load at the other two facilities.
3. Full microgrid: This is the case where power can be fully shared among three facilities.

Please note that in limited and full microgrid mode, when load shedding cannot be avoided, we need to determine how to distribute load shedding among the three facilities. In the following, let the term "net demand" at a facility denote the deficit of local generation in relation to the local load. To ensure realistic and reasonable load shedding, we impose two rules during post processing of the results:

- a facility should not import and export power simultaneously, and

- the amount of energy a facility imports each hour should be proportionate to its net demand. In other words, the total amount of energy export should be proportionally distributed to all facilities requiring energy import according to their net demand.

2.2.3. DG Failure

Three types of DG failure described in [25] are considered in this resiliency analysis:

1. Fail to start: a DG fails to start up on demand.
2. Fail to load: a DG fails to pick up load after started.
3. Fail to run: a DG fails in the second hour of serving load or later.

The status of each DG is simulated in each hour according to the flowchart in Figure 12. The function rand() generates a random number that is uniformly distributed between 0 and 1. PFTS, PFTL, and PFTR respectively denote the probability of the three types of failure in an hour. Once a failure occurs, the corresponding DG becomes unavailable and DERs need to be re-dispatched for the remaining hours of an outage.

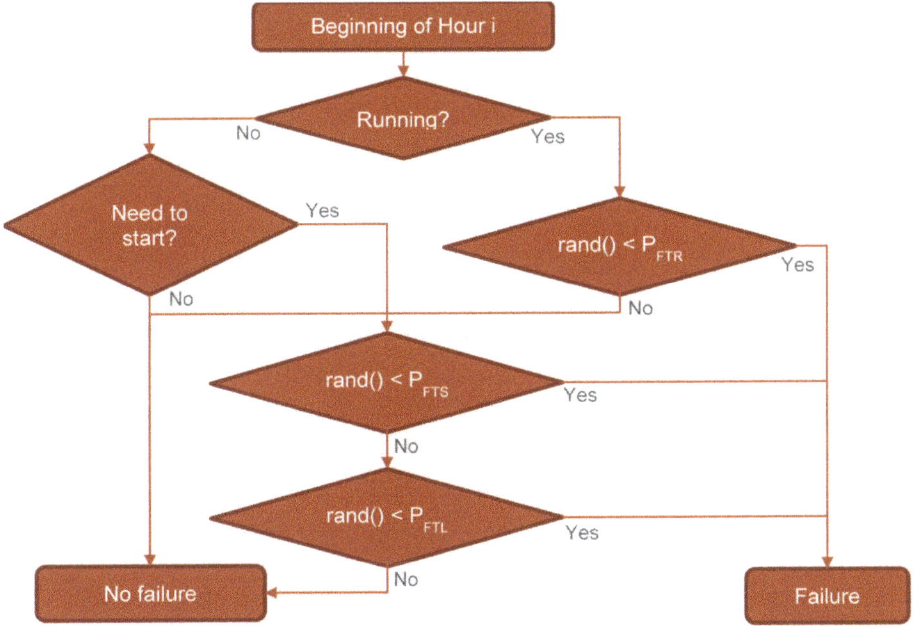

Figure 12. Flowchart to simulate distributed generator (DG) failure in each hour.

The optimal dispatch of available assets is formulated as a linear programming problem. The primary objective is to minimize total unserved energy and the secondary is to minimize fuel consumption. The decision variables include: generation power (negative for BESS charging) from each asset in the k-th hour $p_k^{(\langle asset\_type\rangle,\langle site\rangle)}$, BESS discharging power p_k^+, charging power p_k^-, energy state $s_k^{(\mathrm{CDH})}$, imported power at each site in the k-th hour $p_k^{(i,\langle site\rangle)}$, and unserved load at each site in the k-th hour $\tilde{l}_k^{(\langle site\rangle)}$. Input parameters are hourly load at each site $l_k^{(\langle site\rangle)}$, energy capacity of the BESS at CDH $\mathrm{E}^{(\mathrm{CDH})}$, the efficiency of charging η^- and discharging η^+, rated power of each asset $\mathrm{P}^{(\langle asset\_type\rangle,\langle site\rangle)}$, and hourly power output from PV $\mathrm{P}_k^{(\mathrm{PV},\langle site\rangle)}$ in maximum power point tracking mode.

In the full microgrid mode, the optimization is formulated as minimize:

$$\lambda \sum_{k} \left(\tilde{l}_k^{(\mathrm{CDH})} + \tilde{l}_k^{(\mathrm{SVAHS})} + \tilde{l}_k^{(\mathrm{DPW})} \right) + R \sum_{k} \left(p_k^{(\mathrm{DG,\,CDH})} + p_k^{(\mathrm{DG,\,SVAHS})} + p_k^{(\mathrm{DG,DPW})} \right) \tag{1}$$

subject to

$$0 \le s_k^{(\mathrm{CDH})} \le \mathrm{E}^{(\mathrm{CDH})},\ \forall k, \tag{2}$$

$$p_k^{(\mathrm{BESS,CDH})} = p_k^{+} + p_k^{-}, -\mathrm{P}^{(\mathrm{BESS,CDH})} \le p_k^{-} \le 0 \le p_k^{+} \le \mathrm{P}^{(\mathrm{BESS,CDH})}, \forall k, \tag{3}$$

$$s_k^{(\mathrm{CDH})} - s_{k-1}^{(\mathrm{CDH})} = -\frac{p_k^{+}}{\eta^{+}} - \eta^{-} p_k^{-}, \forall k, \tag{4}$$

$$0 \le p_k^{(\mathrm{DG,CDH})} \le \mathrm{P}^{(\mathrm{DG,CDH})}, \forall k, \tag{5}$$

$$0 \le p_k^{(\mathrm{DG,SVAHS})} \le \mathrm{P}^{(\mathrm{DG,SVAHS})}, \forall k, \tag{6}$$

$$0 \le p_k^{(\mathrm{DG,DPW})} \le \mathrm{P}^{(\mathrm{DG,DPW})}, \forall k, \tag{7}$$

$$0 \le p_k^{(\mathrm{PV,CDH})} \le \mathrm{P}_k^{(\mathrm{PV,CDH})}, \forall k, \tag{8}$$

$$0 \le p_k^{(\mathrm{PV,SVAHS})} \le \mathrm{P}_k^{(\mathrm{PV,SVAHS})}, \forall k, \tag{9}$$

$$\tilde{l}_k^{(\mathrm{CDH})} \ge 0,\ \tilde{l}_k^{(\mathrm{SVAHS})} \ge 0,\ \tilde{l}_k^{(\mathrm{DPW})} \ge 0,\ \forall k, \tag{10}$$

$$p_k^{(\mathrm{BESS,CDH})} + p_k^{(\mathrm{DG,CDH})} + p_k^{(\mathrm{PV,CDH})} + p_k^{(\mathrm{i,CDH})} = l_k^{(\mathrm{CDH})} - \tilde{l}_k^{(\mathrm{CDH})},\ \forall k, \tag{11}$$

$$p_k^{(\mathrm{DG,SVAHS})} + p_k^{(\mathrm{PV,SVAHS})} + p_k^{(\mathrm{i,SVAHS})} = l_k^{(\mathrm{SVAHS})} - \tilde{l}_k^{(\mathrm{SVAHS})},\ \forall k, \tag{12}$$

$$p_k^{(\mathrm{DG,DPW})} + p_k^{(\mathrm{i,DPW})} = l_k^{(\mathrm{DPW})} - \tilde{l}_k^{(\mathrm{DPW})},\ \forall k, \tag{13}$$

$$p_k^{(\mathrm{i,CDH})} + p_k^{(\mathrm{i,SVAHS})} + p_k^{(\mathrm{i,DPW})} = 0,\ \forall k, \tag{14}$$

where R is the fuel consumption rate of DGs and λ is a large constant to guarantee that reducing unserved load energy is prioritized over fuel conservation.

In the limited microgrid mode, only the power from PV and the BESS can be exported to support the load at the other two facilities. Another constraint is added to ensure that the power exported from CDH does not exceed the total power from the BESS and PV:

$$-p_k^{(\mathrm{i,CDH})} \le p_k^{(\mathrm{BESS,CDH})} + p_k^{(\mathrm{PV,CDH})},\ \forall k. \tag{15}$$

In the case of no microgrid, (15) should be replaced by (16) to disable power sharing.

$$p_k^{(\mathrm{i,CDH})} = p_k^{(\mathrm{i,SVAHS})} = p_k^{(\mathrm{i,DPW})} = 0,\ \forall k \tag{16}$$

3. Results

This section is divided into two sections presenting first the economic analysis results and, last, the results of the resiliency analysis.

3.1. Economic Analysis Results

The economic analysis is designed to define the value that the NMP can achieve and to inform the development of operational guidelines for securing value post deployment. In so doing, the analysis could also be useful to other microgrid operators facing similar investment decisions and those attempting to extract maximum value from existing microgrid assets.

3.1.1. Evaluation of Benefits and Costs

In this section, benefits associated with BESS and PV operations are specified. Results are presented in Table 6 and Figure 13.

Table 6. Northampton microgrid 20-year present value benefits vs. costs.

Component	Benefits	Costs
Demand Charge Reduction	$183,662	-
NG Demand Response	$711,674	-
ICAP Tag Reduction	$605,555	-
Outage Mitigation	$274,308	-
PV Energy Payment Reduction	$478,620	-
PV RECs	$256,878	-
BESS Costs	-	$1,026,833
PV Costs	-	$1,138,700
Total Benefits and Costs	$2,510,697	$2,165,533

Figure 13. Twenty-year benefits and costs for the Northampton microgrid.

Over the 20-year economic life of the BESS and PV array, we estimate that the benefits would total approximately $2.5 million, presented in present value terms. Of the benefits tied to BESS operation, the largest portion (28.3% of total benefits) results from demand response benefits, which measure $711.7 thousand. The second largest benefit of BESS operations is from ICAP tag reduction at $605.6 thousand or 24.1% of total benefits. PV benefits, comprised of energy savings and RECs, total $735.5 thousand or 29.3% of the total. When compared to present value system costs of $2.2 million, net benefits are calculated at $345.2 thousand. The overall return on investment is 1.16, meaning that every dollar of investment would be expected to yield $1.16 in returns.

The research team conducted sensitivity analysis (SA) to test the effects of varying certain parameters on study results. The various scenarios are outlined below and their impacts were measured in comparison to the base case. SA was performed by making the following adjustments to the assumptions:

- SA 1: Vary discount rate by +/−1%
- SA 2: Vary price growth rate by +/−1%
- Use DOER grant to eliminate cost of BESS.

The results of each SA are presented in Figure 14. Note that the table with results appears below the figure. Decreasing the discount rate by 1% led to an increase of $311 thousand in total net present value (NPV) benefits, while increasing it by 1% decreased NPV by $166 thousand. Increasing the annual benefit growth rate by 1% resulted in an NPV increase of $264 thousand and a 1% decrease resulted in an NPV drop of $129 thousand. Using a grant from the DOER to eliminate all capital costs associated with the BESS would improve NPV by $577 thousand.

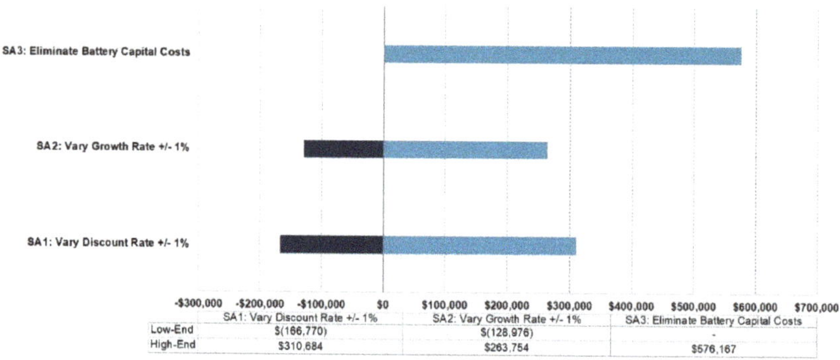

Figure 14. Sensitivity analysis results.

3.1.2. Financial Impact of Incentives

There are a number of incentive programs that could improve the financial results of the NMP. NMP partners have already taken advantage of the Community Clean Energy Resiliency Initiative administered by the Massachusetts DOER. The NMP was awarded a $3.1 million grant under this program, which was designed to address long-duration outages, a need that became clear after Massachusetts experienced significant electricity service disruptions in the aftermath of Hurricane Sandy.

Several clean energy incentive programs could also yield benefits for this and other microgrid projects. If the PV and/or storage were procured through a power purchase agreement with a private third-party vendor, additional tax benefits could accrue to the vendor and ultimately be passed through to the partners. With the capital cost of the PV system and BESS exceeding $1.7 million, the value of a 30% Federal Investment Tax Credit would be $515 thousand. Microgrid financials could also be improved as a result of a statewide carbon tax. Carbon taxes around the world reach as high as $139 per ton CO_2e in Sweden [26]. To explore the value of a $50/ton carbon tax, the annual value of CO_2 savings associated with PV production (406 MWh) was first multiplied by the average CO_2 emissions (808 pounds) per MWh of net generation for the State of Massachusetts [27]. That yields 164 tons of annual CO_2 emission reductions, which would be valued at $8206 annually or $160 thousand in present value terms over a 25-year economic life of the PV array.

3.2. Resiliency Analysis Results

Following the proposed modeling method, we performed comprehensive analysis for the system considering 72 scenarios with different combinations of the following aspects:

1. With and without the 386 kW PV at CDH
2. Outage durations: 3 days, 7 days, and 14 days
3. With and without modeling DG failure rates
4. Seasons: summer (June–October) and winter (November–May)
5. System connectedness: no microgrid, limited microgrid, and full microgrid.

In all scenarios, 10,000 outage events are generated to evaluate the system survivability. Key resiliency analysis results are summarized as follows.

3.2.1. With PV Array at CDH

One of the key metrics of resiliency is the survival rate of a system. It is calculated as

$$\text{Survival Rate} = \frac{\text{Number of outage instances where no load shedding occured}}{\text{Total number of simulated outage instances}} \times 100\%. \quad (17)$$

The survival rate and fuel consumption of the system under various scenarios are presented in Figure 15. In the figure, different colors are used to differentiate facilities, and shapes for system connectedness. The summer season is denoted by filled markers, whereas the winter season is represented by unfilled ones. Marker sizes are used to represent outage lengths. In addition, a dot is added to the center of the marker for scenarios where DG failures are considered. The exact numbers are recorded in tables in the following sections for 3 days, 7 days, and 14 days outages, respectively.

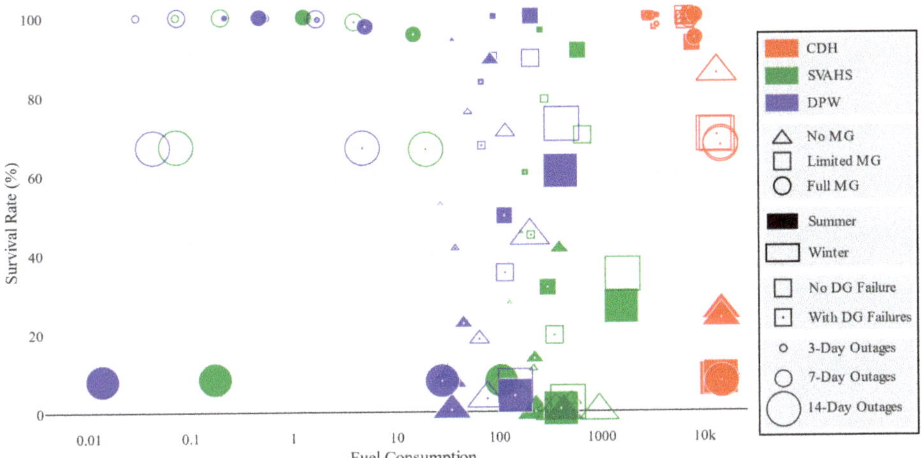

Figure 15. Survival rate and fuel consumption of the system with PV arrays at CDH.

Three Days Outages

The survivability for each facility with different levels of connectedness is summarized in the key observations and insights provided below, which are based on the results presented in Table 7.

Table 7. Mean survival rates of the system with CDH PV in 3 days outages.

Season	Scenario	No DG Failure			With DG Failures		
		CDH	SVAHS	DPW	CDH	SVAHS	DPW
Summer	No MG	100.00	45.39	94.11	96.75	27.58	52.69
	Limited MG	100.00	96.50	100.00	97.06	60.50	83.44
	Full MG	100.00	100.00	100.00	97.67	98.20	99.36
Winter	No MG	100.00	10.92	75.85	99.55	6.54	41.43
	Limited MG	100.00	79.10	89.92	99.57	44.70	67.37
	Full MG	100.00	100.00	100.00	99.64	99.61	99.90

MG = Microgrid.

- Forming a microgrid helps to increase survivability for all facilities, and the "Full Microgrid" mode results in the highest resiliency level.
- In the "No Microgrid" and "Limited Microgrid" modes, SVAHS and DPW are more likely to survive outages in summer than those in winter because their loads are generally lower in summer.

- When DG failures are considered, survivability of CDH in the "Full Microgrid" mode is slightly better than that in the "Limited Microgrid" mode. Limited power-sharing from CDH causes frequent charging and discharging of the battery, which leads to more frequent DG start-ups and increasing DG failures.

The mean total fuel consumption of the system in the same outages are presented in Table 8. The diesel reserve of 15,200 gallons is more than enough to cope with 3 days outages. In fact, it is likely to be sufficient for any outage shorter than two weeks.

Table 8. Mean diesel fuel consumption (gallons) of the system with CDH PV in 3 days outages.

Season	Scenario	No DG Failure			With DG Failures		
		CDH	SVAHS	DPW	CDH	SVAHS	DPW
Summer	No MG	3385	166	36	3379	127	27
	Limited MG	3243	260	90	3321	184	68
	Full MG	3592	1	0	3583	5	2
Winter	No MG	2886	218	51	2883	165	38
	Limited MG	2802	287	90	2861	208	68
	Full MG	3179	0	0	3177	1	1

MG = Microgrid.

Seven Days Outages

The survival rates of the system and the fuel consumption in 7 days outages are summarized in Tables 9 and 10, respectively. We have very similar observations as in 3 days outages. The benefit of forming a microgrid becomes more significant.

Table 9. Mean survival rates of the system with CDH PV in 7 days outages.

Season	Scenario	No DG Failure			With DG Failures		
		CDH	SVAHS	DPW	CDH	SVAHS	DPW
Summer	No MG	100.00	41.24	88.95	94.39	13.51	22.42
	Limited MG	100.00	91.29	100.00	93.03	31.38	49.61
	Full MG	100.00	100.00	100.00	94.38	95.56	97.54
Winter	No MG	100.00	2.91	70.75	98.68	0.92	18.43
	Limited MG	100.00	69.86	89.35	98.76	19.34	35.17
	Full MG	100.00	100.00	100.00	98.99	98.72	99.57

MG = Microgrid.

Table 10. Mean diesel fuel consumption (gallons) of the system with CDH PV in 7 days outages.

Season	Scenario	No DG Failure			With DG Failures		
		CDH	SVAHS	DPW	CDH	SVAHS	DPW
Summer	No MG	7889	391	84	7874	224	45
	Limited MG	7569	600	210	7854	302	114
	Full MG	8378	1	0	8346	15	5
Winter	No MG	6758	506	118	6754	286	64
	Limited MG	6564	665	210	6783	349	115
	Full MG	7440	0	0	7433	4	2

MG = Microgrid.

Fourteen Days Outages

The survival rates of the system and the fuel consumption in 14 days outages are summarized in Tables 11 and 12, respectively. Key observations and insights are provided below.

Table 11. Mean survival rates of the system with CDH PV in 14 days outages.

Season	Scenario	No DG Failure			With DG Failures		
		CDH	SVAHS	DPW	CDH	SVAHS	DPW
Summer	No MG	25.27	0.08	8.35	23.50	0.03	0.49
	Limited MG	8.02	26.53	60.86	8.41	0.76	4.04
	Full MG	8.04	8.39	7.98	7.51	7.83	7.92
Winter	No MG	85.50	0.00	44.43	85.47	0.00	3.29
	Limited MG	70.04	34.63	72.74	69.78	2.38	6.54
	Full MG	67.71	67.29	67.12	67.18	66.48	66.88

MG = Microgrid.

Table 12. Mean diesel fuel consumption (gallons) of the system with CDH PV in 14 days outages.

Season	Scenario	No DG Failure			With DG Failures		
		CDH	SVAHS	DPW	CDH	SVAHS	DPW
Summer	No MG	14,713	428	31	14,696	230	34
	Limited MG	13,195	1573	405	14,462	401	143
	Full MG	15,173	0	0	14,775	105	28
Winter	No MG	13,475	945	203	13,465	357	76
	Limited MG	12,672	1622	418	13,548	468	144
	Full MG	14,712	0	0	14,647	20	5

MG = Microgrid.

- Compared with 3 days and 7 days outages, survival rates drop significantly for prolonged outages due to fuel shortages;
- Survival rates in summer are lower because the total energy consumption is higher;
- CDH is more likely to survive in the "No Microgrid" mode, because fuel is conserved for CDH by not sharing power while shedding loads at SVAHS and DPW.

3.2.2. Without PV Array at CDH

To evaluate the benefits of installing the PV array at CDH in terms of the resiliency improvement of the system, we run a similar analysis for the system without the PV array and compare the results with the previous ones. The results are presented in Figure 16.

Three Days Outages

Compared with the case with CDH PV, significantly lower survival rates of SVAHS and DPW under the "Limited Microgrid" mode are observed because the PV and BESS were the only assets that can help support the load at SVAHS and DPW under this mode. Results are presented in Tables 13 and 14.

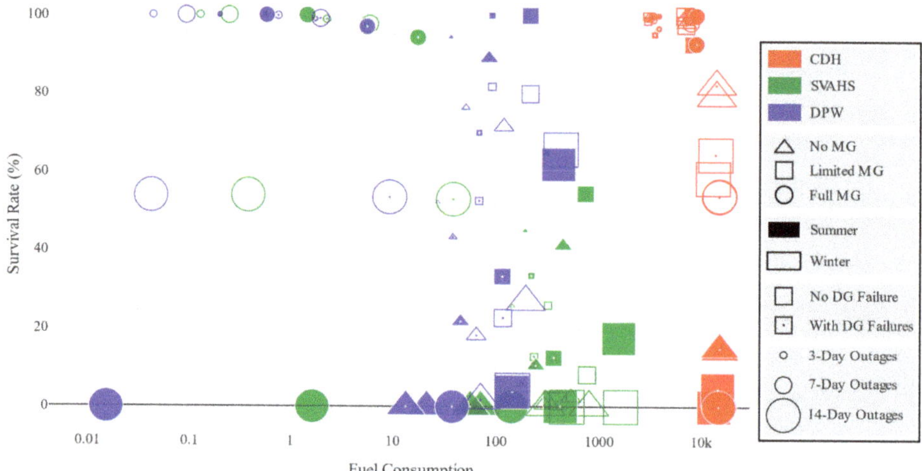

Figure 16. Survival rate and fuel consumption of the system without PV arrays at CDH.

Table 13. Mean Survival rates of the system without CDH PV in 3 days outages.

Season	Scenario	No DG Failure			With DG Failures		
		CDH	SVAHS	DPW	CDH	SVAHS	DPW
Summer	No MG	100.00	44.97	94.49	96.20	25.80	52.37
	Limited MG	100.00	64.94	100.00	95.24	33.53	70.05
	Full MG	100.00	100.00	100.00	96.81	97.59	99.03
Winter	No MG	100.00	10.52	76.48	99.09	5.75	43.36
	Limited MG	100.00	26.01	81.79	98.73	12.79	52.70
	Full MG	100.00	100.00	100.00	99.11	98.94	99.87

MG = Microgrid.

Table 14. Mean diesel fuel consumption (gallons) of the system without CDH PV in 3 days outages.

Season	Scenario	No DG Failure			With DG Failures		
		CDH	SVAHS	DPW	CDH	SVAHS	DPW
Summer	No MG	3515	189	36	3507	143	27
	Limited MG	3363	306	90	3422	218	68
	Full MG	3761	1	0	3749	5	2
Winter	No MG	2974	228	50	2971	172	38
	Limited MG	2874	318	90	2918	233	68
	Full MG	3291	0	0	3287	2	1

MG = Microgrid.

Seven Days Outages

For 7 days outages, there are similar observations as for 3 days outages. The survivability differences between "No DG Failure" and "With DG Failures", as well as that among different levels of system connectedness, become more significant due to the longer outage duration. Results are presented in Tables 15 and 16.

Table 15. Mean survival rates of the system without CDH PV in 7 days outages.

Season	Scenario	No DG Failure			With DG Failures		
		CDH	SVAHS	DPW	CDH	SVAHS	DPW
Summer	No MG	100.00	41.19	89.13	93.34	10.51	21.83
	Limited MG	100.00	54.54	100.00	92.76	12.53	33.30
	Full MG	100.00	100.00	100.00	92.85	94.37	97.10
Winter	No MG	100.00	2.48	71.64	98.17	0.57	18.28
	Limited MG	100.00	8.18	79.96	97.68	1.52	22.73
	Full MG	100.00	100.00	100.00	98.32	97.82	99.16

Table 16. Mean diesel fuel consumption (gallons) of the system without CDH PV in 7 days outages.

Season	Scenario	No DG Failure			With DG Failures		
		CDH	SVAHS	DPW	CDH	SVAHS	DPW
Summer	No MG	8203	443	84	8182	242	45
	Limited MG	7836	728	210	8062	362	114
	Full MG	8778	1	1	8738	17	5
Winter	No MG	6956	529	117	6952	291	64
	Limited MG	6703	760	210	6873	398	116
	Full MG	7693	0	0	7682	6	2

MG = Microgrid.

Fourteen Days Outages

Compared to results with PV arrays at CDH included, worse survival rates across the board are observed. The energy that would have been generated by PV arrays will have to be generated by DGs instead. The fuel shortage is therefore aggravated. Results are presented in Tables 17 and 18.

Table 17. Mean survival rates of the system without CDH PV in 14 days outages.

Season	Scenario	No DG Failure			With DG Failures		
		CDH	SVAHS	DPW	CDH	SVAHS	DPW
Summer	No MG	14.17	0.00	0.00	14.96	0.00	0.01
	Limited MG	0.00	17.56	62.03	4.48	0.09	3.81
	Full MG	0.00	0.00	0.00	0.00	0.00	0.00
Winter	No MG	79.22	0.00	26.93	82.20	0.00	1.67
	Limited MG	58.47	0.00	65.79	64.49	0.00	4.24
	Full MG	54.02	54.04	53.98	53.81	53.06	53.59

MG = Microgrid.

Table 18. Mean diesel fuel consumption (Gallons) of the system without CDH PV in 14 days outages.

Season	Scenario	No DG Failure			With DG Failures		
		CDH	SVAHS	DPW	CDH	SVAHS	DPW
Summer	No MG	15,107	71	21	15,058	57	13
	Limited MG	13,253	1549	398	14,451	418	143
	Full MG	15,198	2	0	14,748	140	38
Winter	No MG	13,839	795	192	13,822	302	71
	Limited MG	12,866	1621	418	13,598	487	144
	Full MG	14,928	0	0	14,800	38	9

MG = Microgrid.

An interesting observation under the "No Microgrid" mode is that CDH is more likely to survive when DG failures are considered. This is because the failures of DGs at SVAHS and DPW effectively save fuel for CDH. The benefit of such savings is manifested in the 14 days outages cases due to the shortage of fuel. Similar phenomena can also be observed in the "Limited Microgrid" mode.

4. Discussion

This report examines the economic and resilience benefits of a proposed microgrid in Northampton, Massachusetts that would link the Northampton DPW, CDH, and SVAHS. We modeled financial benefits of microgrid operations—including demand response revenue, capacity charge reduction, demand charge reduction, outage mitigation, energy charge reductions due to solar PV energy production, and renewable energy credits—by simulating one-year of microgrid operations in BSET. The simulation resulted in an estimate that is technically achievable given the capacities of the microgrid assets and the co-optimization algorithms that ensure no double counting of benefits. To evaluate resilience benefits, we performed a comprehensive analysis for the system considering 72 scenarios characterized by changes in outage duration, season, system configuration, and distributed generator availability.

There are a number of studies that have estimated the economic benefits of DERs in the northeastern U.S. In these studies, outage mitigation benefits are typically measured in terms of improvements in outage statistics (e.g., system average interruption duration index, system average interruption frequency index) or VOLL to customers. In [19], the Nantucket Island distribution network was modeled using two open-source simulation programs, OpenDSS (Electric Power Research Institute, Palo Alto, CA, USA) and GridLAB-D (PNNL, Richland, WA, USA), and historic outages over an 11-year period were modeled with the change in outage costs to customers due to the use of a combustion turbine generator with a temperature-dependence maximum capacity of 16 MW and a 6 MW/48 MWh lithium-ion BESS being measured. In [28], VOLL was estimated for a number of scenarios varying based on BESS energy capacity, with the BESS being used to provide power to the town's police station and dispatch center in an islanded mode during outages.

This paper presents a comprehensive approach that considers both economic and resilience benefits. While the economic analysis effectively captures stacked value streams and an innovative method was used to quantify microgrid survivability against a random outage, more research could be done to evaluate tradeoffs between economic and resilience objectives. In addition, a refined model could be used to optimally scale microgrid assets given resilience targets or to maximize resilience benefits with net costs of the assets set to meet certain budget constraints.

5. Conclusions

This report examined the operational and economic benefits of a microgrid proposed in Northampton, Massachusetts that would link the Northampton DPW, CDH, and SVAHS. Historically, all three of these facilities have experienced half-day to two days outages due to severe weather events with the longest outage in recent years lasting three days. A DOER-supported analysis investigated the potential to link these three facilities together and confirmed the capability of the three campuses to island from the grid and continue operating for up to three days with grant-supported investments in microgrid infrastructure. If successfully implemented, the microgrid could offer substantial benefits to the facilities by mitigating the outages to which they have historically been subjected.

PNNL was engaged by the U.S. Department of Energy to review, modify, and model these potential benefits, as well as evaluate the technical performance and financial opportunities available to the NMP partners. The results provide insights into operation of the microgrid by the three partners. The following lessons were drawn from this analysis:

1. Over a 20-year life, the BESS and the 386 kW solar array are estimated to generate $2.5 million in present value benefits. This value exceeds the $2.2 million in present value costs necessary to

install, maintain, and operate the assets. The overall ROI ratio for the project under this scenario is 1.16.

2. Sensitivity analysis showed that decreasing the discount rate by one percent led to an increase of $311 thousand in NPV benefits while increasing it by one percent decreased NPV by $166 thousand. Increasing the annual benefit growth rate by 1% resulted in an NPV increase of $264 thousand and a 1% decrease resulted in an NPV drop of $129 thousand. Using a grant from the DOER to eliminate all capital costs associated with the BESS would improve NPV by $577 thousand.
3. With the PV array included at CDH, forming a microgrid helps increase survivability of all facilities when a three days outage strikes. Limited power sharing under this scenario causes frequent charging and discharging of the battery system, which leads to more frequent DG start-ups and increasing DG failures. When there is a full microgrid, no DG failure, and it is the winter season, all three facilities have a survivability of 100%. With no microgrid and the assumption of DG failures, the SVAHS survivability drops to 6.54% and DPW survivability drops to 41.43%.
4. Under seven days outage scenarios, the benefits of forming a microgrid become more significant. With full sharing between all microgrid members, all facilities are able to withstand 100% of the outage in both summer and winter. When factoring in DG failure probability, the average survivability across all three facilities when the microgrid exists drops to around 95% per facility in summer and 99% per facility in winter. If no microgrid exists, the survivability of SVAHS and the DPW during a seven days outage in summer and assuming no DG failure drops to 41.24% and 88.95%, respectively. Under the same scenario, but with DG failure assumed, the SVAHS survivability drops to <1% and DPW to 18.43%.
5. When 14 days outages are considered, survivability drops significantly due to fuel shortage. Summer survivability is lower than winter due to higher total energy consumption. Under all scenarios under this outage duration, CDH has higher survivability when no microgrid exists. When a full microgrid is assumed, all facilities have an approximate eight percent survivability rate in summer regardless of DG failure. In winter, this increases substantially to about 67% across all facilities. In summer and without a microgrid, CDH has an approximate 25% survivability rate while SVAHS has <1% and DPW has an 8.35% survivability, assuming no DG failure. If DG failure is assumed, both SVAHS and DPW have <1% chance of surviving the outage while CDH only drops to 23.5%.
6. When the potential PV array at CDH is removed from the microgrid configuration, survival rates drop for SVAHS and DPW especially. Compared with the case with CDH and PV, significantly lower survival rates of SVAHS and DPW under "Limited Microgrid" mode are observed, because the PV and BESS were the only assets that can help support the load at SVAHS and DPW under this mode.
7. The worst subset of scenarios for resilience across all scenarios is a 14 days outage with no PV. Under this configuration and when there is a full microgrid in summer, there is a zero percent survivability rate across all facilities both with and without DG failure considered. In winter, the survivability under the same configuration increases to approximately 53% for each facility. CDH is consistently better off with no microgrid during long-duration outages as they are able to reserve fuel that would otherwise be shared with other microgrid members.

The results of this and other similar microgrid assessments demonstrate that when certain conditions are present, well-designed microgrids hold the potential to achieve resiliency goals while largely, if not entirely, paying for themselves through grid operations. While diesel and natural gas-fired generators may offer the least costly option for achieving resiliency goals, their operation may be limited to emergency conditions due to noise and emission constraints. Microgrids that include generators plus PV and BESSs could lower the net cost of the system required to meet resiliency goals by generating revenue during normal grid operations to offset system costs.

Author Contributions: Conceptualization, P.B., D.W. (Di Wu), and R.D.; validation, P.B.; formal analysis, K.M., P.B., D.W. (Di Wu), D.W. (Dexin Wang) and R.D.; investigation, K.M., P.B., D.W. (Di Wu), D.W. (Dexin Wang), R.D. and V.F.; resources, P.B.; data curation, K.M., P.B., D.W. (Di Wu), D.W. (Dexin Wang), R.D. and V.F.; writing—original draft preparation K.M., P.B., D.W. (Di Wu), D.W. (Dexin Wang) and R.D.; writing—review and editing, P.B.; visualization, P.B., D.W. (Di Wu) and R.D.; supervision, P.B.; project administration, P.B.; funding acquisition, P.B. All authors have read and agreed to the published version of the manuscript.

Funding: This research was funded by the U.S. Department of Energy under contract number DE-AC05-76RL01830.

Acknowledgments: We are grateful to Imre Gyuk, who is the Energy Storage Program Manager in the Office of Electricity at the U.S. Department of Energy, for providing financial support and leadership on this and other related work at Pacific Northwest National Laboratory. We are also grateful to Chris Mason of the City of Northampton, Massachusetts for the data, information, and insights he offered the team throughout the study.

Conflicts of Interest: The authors declare no conflict of interest.

Abbreviations

BESS	battery energy storage system
BSET	Battery Storage Evaluation Tool
BTM	behind-the-meter
CDH	Cooley Dickenson Hospital
DER	distributed energy resources
DG	distributed generator
DOER	Massachusetts Department of Energy Resources
DPW	Northampton Department of Public Works
FEDS	Federal Energy Decisions System
HVAC	heating, ventilation, and air conditioning
ICAP	installed capacity
ISO-NE	Independent System Operator New England
NMP	Northampton Microgrid Project
NPV	net present value
NSRDB	National Solar Radiation Database
O&M	operations and maintenance
PNNL	Pacific Northwest National Laboratory
PSM	physical solar model
PV	photovoltaics
REC	renewable energy credit
ROI	return on investment
SA	sensitivity analysis
SVAHS	Smith Vocational Area High School
VOLL	value of lost load

References

1. Balducci, P.; Alam, M.J.E.; Hardy, T.D.; Wu, D. Assigning value to energy storage systems at multiple points in an electrical grid. *Energy Environ. Sci.* **2018**, *11*, 1926–1944. [CrossRef]
2. Parhizi, S.; Lotfi, J.; Khodaei, A.; Bahramirad, S. State of the art in research on microgrids: A review. *IEEE Access* **2015**, *3*, 890–925. [CrossRef]
3. Bahramara, S.; Sheikhahmadi, P.; Golpîra, H. Co-optimization of energy and reserve in standalone micro-grid considering uncertainties. *Energy* **2019**, *176*, 792–804. [CrossRef]
4. Hajipour, E.; Bozorg, M.; Fotuhi-Firuzabad, M. Stochastic capacity expansion planning of remote microgrids with wind farms and energy storage. *IEEE Trans. Sustain. Energy* **2015**, *6*, 491–498. [CrossRef]
5. Yuan, C.; Illindala, M.S.; Khalsa, A.S. Co-optimization scheme for distributed energy resource planning in community microgrids. *IEEE Trans. Sustain. Energy* **2017**, *8*, 1351–1360. [CrossRef]
6. Babazadeh, H.; Gao, W.; Wu, Z.; Li, Y. Optimal energy management of wind power generation system in islanded microgrid system. In Proceedings of the North American Power Symposium, Manhattan, KS, USA, 22–24 September 2013; pp. 1–5. [CrossRef]
7. Alhaider, M.; Fan, L. Mixed integer programming based battery sizing. *Energy Syst.* **2014**, *5*, 787–805. [CrossRef]

8. Yuan, C.; Illindala, M.S.; Haj-ahmed, M.A.; Khalsa, A.S. Distributed energy resource planning for microgrids in the United States. In Proceedings of the IEEE Industry Applications Society Annual Meeting, Addison, TX, USA, 18–22 October 2015; pp. 1–9. [CrossRef]
9. Fotedar, V.; Balducci, P.; Warwick, M.; Wu, D.; Wang, D.; Mongird, K. *Opportunities for Dispatchable Power Projects in the New England System Operator Area*; PNNL-29279; Pacific Northwest National Laboratory: Richland, WA, USA, 2019.
10. Khodaei, A. Resiliency-oriented microgrid optimal scheduling. *IEEE Trans. Smart Grid* **2014**, *5*, 1584–1591. [CrossRef]
11. Rosales-Asensio, E.; de Simón-Martín, M.; Borge-Diez, D.; Blanes-Peiró, J.; Colmenar-Santos, A. Microgrids with energy storage systems as a means to increase power resilience: An application to office buildings. *Energy* **2019**, *172*, 1005–1015. [CrossRef]
12. Simpkins, T.; Anderson, K.; Cutler, D.; Olis, D. Optimal sizing of a solar-plus-storage system for utility bill savings and resiliency benefits. In Proceedings of the Innovative Smart Grid Technologies Conference, Minneapolis, MN, USA, 6–9 September 2016; pp. 1–6. [CrossRef]
13. Laws, N.D.; Anderson, K.; DiOrio, N.A.; Li, X.; McLaren, J. Impacts of valuing resilience on cost optimal PV and storage systems for commercial buildings. *Renew. Energy* **2018**, *127*, 896–909. [CrossRef]
14. Apache Software Foundation. Apache License Version 2.0. Available online: https://www.apache.org/licenses/LICENSE-2.0 (accessed on 11 September 2020).
15. Apache Software Foundation. Google Earth Satellite Imagery. Coordinates: 42.3335308° N, 72.6574563° W. Available online: https://www.google.com/maps/@42.3335308,-72.6574563,144m/data=!3m1!1e3 (accessed on 11 September 2020).
16. Wu, D.; Jin, C.; Balducci, P.; Kintner-Meyer, M. An Energy Storage Assessment: Using Optimal Control Strategies to Capture Multiple Services. In Proceedings of the IEEE Power and Energy Society General Meetings, Denver, CO, USA, 26–30 July 2015; PNNL-SA-106768; IEEE: Piscataway, NJ, USA, 2015; pp. 1–5. [CrossRef]
17. Dahowski, R.T. *Facility Energy Decision System, Release 8.0*; PNNL-29974; Pacific Northwest National Laboratory: Richland, WA, USA, 2020.
18. Apache Software Foundation. Google Earth Satellite Imagery. Coordinates: 42.3313872° N, 72.6552245° W. Available online: https://www.google.com/maps/@42.3313872,-72.6552245,679m/data=!3m1!1e3 (accessed on 11 September 2020).
19. Balducci, P.J.; Alam, M.E.; McDermott, T.E.; Fotedar, V.; Ma, X.; Wu, D.; Bhatti, B.A.; Mongrid, K.; Bhattarai, B.; Crawford, A.J.; et al. *Nantucket Island Energy Storage System Assessment*; PNNL-28941; Pacific Northwest National Laboratory: Richland, WA, USA, 2019.
20. Sullivan, M.; Schellenberg, K.; Blundell, M. *Updated Value of Service Reliability Estimates for Electric Utility Customers in the United States*; Prepared for U.S. Department of Energy by Lawrence Berkeley National Laboratory; Lawrence Berkeley National Laboratory: San Francisco, CA, USA, 2015.
21. "NSRDB Data Viewer". National Renewable Energy Laboratory. Golden, CO. Available online: https://maps.nrel.gov/nsrdb-viewer/?aL=x8CI3i%255Bv%255D%3Dt%26ozt_aP%255Bv%255D%3Dt%26ozt_aP%255Bd%255D%3D1&bL=dI6joO&cE=0&lR=0&mC=42.326189357961205%2C-72.64185905456543&zL=13 (accessed on 15 May 2020).
22. Habte, A.; Sengupta, M.; Lopez, A. *Evaluation of the National Solar Radiation Database (NSRDB) Version 2*; Technical Report NREL/TP-5D00-67722; National Renewable Energy Laboratory: Golden, CO, USA, 2017; pp. 1998–2015.
23. "New England Class I REC Market Update". Power Advisory LLC. Available online: https://poweradvisoryllc.com/new-england-class-i-rec-market-update/ (accessed on 11 May 2020).
24. MassCEC's Solar Costs Comparison Tool. Massachusetts Clean Energy Center. Boston, MA. Available online: https://www.masscec.com/solar-costs-performance (accessed on 18 May 2020).
25. Nuclear Energy Institute. *Regulatory Assessment Performance Indicator Guideline*; Appendix F; Nuclear Energy Institute: Washington, DC, USA, 2013.
26. Balducci, P.; Wu, D.; Ramachandran, T.; Campbell, A.; Fotedar, V.; Mongird, K.; Huang, S.; Bhatnagar, D.; Jones, S.; Smith, C.; et al. *Power-to-Gas. System Valuation: Final Report*; PNNL-ACT-10095; Pacific Northwest National Laboratory: Richland, WA, USA, 2020.

27. Energy Information Administration. State Electricity Profiles, Massachusetts Electricity Profile 2018. Available online: https://www.eia.gov/electricity/state/massachusetts/ (accessed on 1 September 2020).
28. Byrne, R.; Hamilton, S.; Borneo, D.; Olinsky-Paul, T.; Gyuk, I. *The Value Proposition of Energy Storage at the Sterling Municipal Light Department*; SAND2017-1093; Sandia National Laboratories: Albuquerque, NM, USA, 2017.

Article

Optimum Resilient Operation and Control DC Microgrid Based Electric Vehicles Charging Station Powered by Renewable Energy Sources

Khairy Sayed [1], Ahmed G. Abo-Khalil [2,3,*] and Ali S. Alghamdi [2]

1 Faculty of Engineering, Sohag University, Sohag 82524, Egypt; khairy_fathy@yahoo.ca
2 Department of Electrical Engineering, College of Engineering, Majmaah University, Almajmaah 11952, Saudi Arabia; aalghamdi@mu.edu.sa
3 Department of Electrical Engineering, College of Engineering, Assuit University, Assuit 71515, Egypt
* Correspondence: a.abokhalil@mu.edu.sa

Received: 25 September 2019; Accepted: 30 October 2019; Published: 7 November 2019

Abstract: This paper introduces an energy management and control method for DC microgrid supplying electric vehicles (EV) charging station. An Energy Management System (EMS) is developed to manage and control power flow from renewable energy sources to EVs through DC microgrid. An integrated approach for controlling DC microgrid based charging station powered by intermittent renewable energies. A wind turbine (WT) and solar photovoltaic (PV) arrays are integrated into the studied DC microgrid to replace energy from fossil fuel and decrease pollution from carbon emissions. Due to the intermittency of solar and wind generation, the output powers of PV and WT are not guaranteed. For this reason, the capacities of WT, solar PV panels, and the battery system are considered decision parameters to be optimized. The optimized design of the renewable energy system is done to ensure sufficient electricity supply to the EV charging station. Moreover, various renewable energy technologies for supplying EV charging stations to improve their performance are investigated. To evaluate the performance of the used control strategies, simulation is carried out in MATLAB/SIMULINK.

Keywords: DC microgrid; electric vehicles; resilient microgrid; solar PV; wind turbine; charging station; control

1. Introduction

Recently, electric vehicles (EVs) became more widespread, and thus, installing EV charger stations is substantial to satisfy the electrical energy demand of a large number of EVs [1]. However, due to the extended electrical grids, fast EV charger stations, parking lots, and residential areas can supply the electrical energy desired to charge EVs. Energy management control strategies are required for the charger stations for designing and calculating an optimal contracted ability to promote performance and operation [2–4]. Effective battery chargers represent a substantial role in the evolution of modern EVs. The characteristics of the battery charger affect the battery life and charging energy efficiency, as well as the charging time. EV battery chargers should have the key advantages of higher power density, higher efficiency, better reliability, smaller size, lighter weight, and cheaper cost. The operation of the charger circuit relies fundamentally on power circuit topology, power circuit passive and active devices, soft switching techniques and control schemes [5]. Mostly, the EV-charger control techniques can be carried out by using analog/digital controllers, digital signal processors, microcontrollers, and some particular integrated circuits. However, this depends upon the complexity of the power circuit topology, cost, and the power rating of converters. Although plug-in electric vehicles (PEVs) are being promoted in the market with the objective of reducing the pollution from conventional automobiles, the energy demands for charging the EV batteries are still supplied by power generated by conventional fossil fuel sources. For this reason, many researchers have proposed the solution of charging PEVs using renewable energy sources like photovoltaic (PV) and wind. Numerous pilot projects are also carried out to charge PEVs from solar PV and wind energy systems [4–9]. These projects are still in the development stage [10]. Moreover, due to the economic and social benefits, research work on charging stations powered by the PV system has engaged researchers worldwide. Generally, the use of solar energy charger is a dependable source for charging small scale electric vehicles, such as scooters, golf carts, and airport utility carriages [11]. The use of photovoltaic powered chargers in a parking lot is analyzed in [12]. A photovoltaic PV-based charging station that is connected with the utility grid is described in [13,14]. Solar PV parking lot chargers and other application models to supply PEVs with solar energy are explored in [15]. Economic studies of PV powered charging stations have been done by [16,17]. Reference [18] depicts how intelligent control algorithms can support PEVs and PV to integrate with the existing electrical power systems. PV system provides a potential source for PEV of median generation capacity, while PEVs represent a dispatchable load for low and extra PV generation during periods of light load demand.

For the stand-alone microgrid, energy management system (EMS) can control demand/supply balance and maximize the environmental or economic benefits. EMS is a key technology for stable microgrid operation. For the stand-alone wind-diesel microgrid in [19], an optimal EMS strategy is proposed, which optimizes the charging/discharging cycles of storage system and system operation cost according to the prediction of wind turbine (WT) output and load demand. A novel EMS-based on a rolling horizon strategy for a renewable-based microgrid, which includes PV, WT generator, diesel generator, and energy storage system (ESS), is proposed in [20]. In [21], a control strategy to reduce power fluctuations is proposed, which utilizes the ESS to smooth the output power of the wind farm. DC microgrids have a less complex control strategy which only adopts P-V droop, mitigates the need for reactive power compensation, and reduce the circulating reactive power. Furthermore, elimination of frequency and phase angle would ease the resynchronization to the utility grid. Without reactive power and harmonics, DC microgrids could also offer a better quality of power [22]. They can feed the DC loads directly by avoiding the conversion losses.

Renewable energy-based charging stations (wind and solar) are friendly EV charging that reduces fossil fuel exhaustion, optimizes investment cost and accommodates fluctuations of generated power by renewable sources. The evaluation objectives of charging stations include operational performance and customer acceptance of charging equipment; pricing criteria to encourage off-peak charging; and grid impacts [23]. Hence, adopting a DC microgrid is presented for enhancing the resilience and optimum operation of microgrid, including distributed generation [24]. The broad problem considered in this research is the optimization of energy flows in the DC microgrid. For this reason, a stable, robust and optimal supervisory control algorithm is substantial for the large scale hybrid dynamical system of PEV charging station. Since the system is subjected to random variations in solar power and the connected vehicles in the parking lot, the system operation must be robust against these disturbances. In a DC-microgrid, buses can be classified into four types: Generation bus, DC load bus, batteries energy storage system (BESS) bus, and connection bus to AC-microgrid using voltage-source converters (VSCs). Moreover, these types of buses can be divided into two groups according to their contribution to microgrid operation and control, which are power bus and slack bus. The power bus absorbs power from/to the microgrid on its own. Typical examples are variable DC-loads and variable (non-dispatchable) generation, such as photovoltaic and wind turbines generation systems. In contrast, slack buses are responsible for balancing the power surplus/deficit resulting from power buses and maintaining stable operation of the microgrid.

Generally, in the largely inhibited parking lot, interventions are focused on removing fully charged EVs, to give non-charged EVs a chance to be an effective to realize powerful utilization of the charging infrastructure [25]. Specifically, two resilience measures are considered, the resilience related to the amount of energy delivered to EVs and the resilience related to the average charging time, are provided for the EV charging station. Resilience can be defined as the ability of the studied system to withstand disturbance state and return to a regular state quickly. It has become a new challenge facing the EV charging station design [26]. A related key performance indicator (KPI) is the percent of sessions with a low charging time ratio to the total charging time divided by the amount of total connection time [27,28]. However, these sessions are often a burden during peak hours and daytime. To increase availability for other EV users, the fully charged EVs should be removed.

References [29,30] propose power management strategies for an autonomous DC-microgrid based on a PV source, a supercapacitor, electrochemical storage, and a diesel generator. However, these papers have difficulties in achieving power balance, while accounting for the slow start-up characteristic of the diesel generator, the self-discharge of an SC. Moreover, the economic operating mode of the diesel generator can increase the total energy cost of the DC-microgrid. References [31,32] has studied a test bed to investigate the dynamic response of a DC-microgrid to major disturbances, but it did not calculate the resilience of the studied DC-microgrid. References [33,34] have investigated the dynamic response of microgrids powered by renewable energy sources, but they did not define resilience of the studied systems.

This paper presents the control strategy of an isolated standalone EV chargers station incorporated in a DC microgrid. This control strategy is investigated using proportional-integral controllers (PI). This controller will regulate the charging of EVs. The proposed EMS is considered promising, due to its robustness and simplicity that makes this suitable for applications in the future smart DC microgrid. A new resilience measurement is defined as the ratio of the normalized system, integrated within its maximum permissible recovery time after the disturbance to the performance integral in the ordinary state. This measure enables the resilience of various systems to be compared on the same comparative scale. To estimate the resilience of DC microgrid, a resilience measurement scheme is developed.

2. System Description

A DC microgrid is a low-voltage network that consists of several energy components, such as controllable loads and distributed energy resources (DERs). The standalone system can decrease the carbon footprint and reduce the losses of power transmission [35]. Figure 1 shows a standalone DC microgrid supplying EVs charging station. The studied system is composed of WT, photovoltaic (PV), and energy storage systems (ESS), such as battery bank. In this system, controllable loads include electric vehicles (EVs). The EVs are charged from DC microgrid through DC-DC converters controlled by a charging regulation control scheme. The battery bank has a dual power flow in the whole system that acts as the energy provider and consumer according to the condition of wind turbine and PV panels' production. The configuration of the standalone charging station is shown in Figure 1. The PV connected to the DC microgrid through a DC-DC converter controlled by the maximum power point tracker (MPPT) scheme. The wind generator connected to the DC microgrid through an AC-DC and DC-DC converters. At the same time, the battery bank charged and discharged from DC microgrid using a bidirectional DC-DC converter.

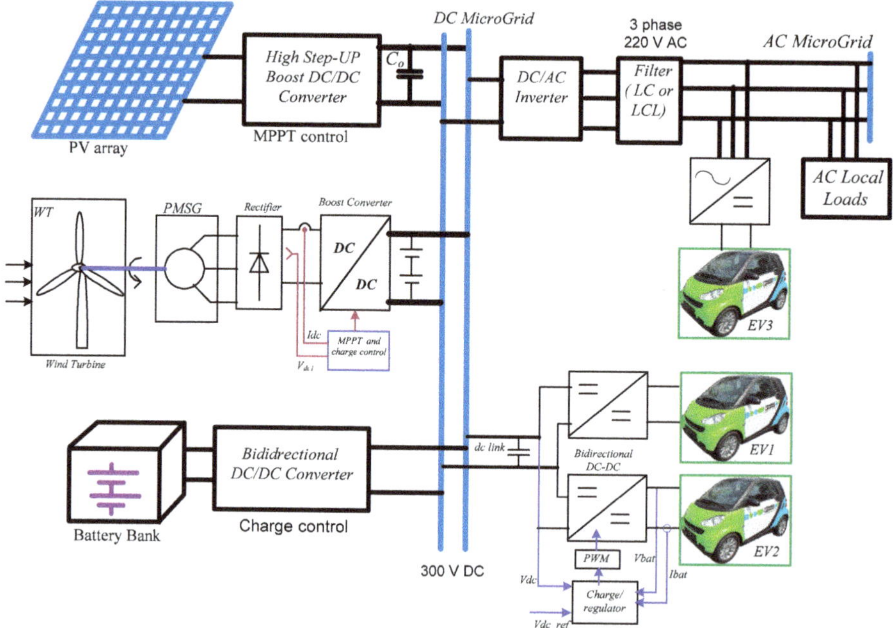

Figure 1. Configuration of the proposed DC microgrid based charging station.

2.1. Configuration of Electric Vehicle

The PEV has three major subsystems: Electric propulsion subsystem, Power source, and Auxiliary system. The electrical propulsion subsystem is composed of the electronic controller; DC-AC power converter; electric traction motor; driving wheels, and mechanical transmission. Based on the control input signals from the accelerator and brake pedals, the electronic or digital controller provides the required control commands to switch on or off the power converters which in turn coordinate the power flow between the electric motor and the EV battery source. However, the backward power flow is due to regenerative braking of the EV, and consequent regenerative energy can be maintained to charge EV Battery. The energy management unit (EMU) cooperates with the electronic controller to deal with regenerative braking and its recovered energy. Generally, the auxiliary power system supplies the necessary energy at various voltage levels to auxiliaries in EV, particularly the power steering units and temperature control. The successful utilization of EVs over the next decade is promoted according to international standards and regulations. Safety codes and standards define a wide range of issues related to EVs. For example, article 625-18 of the national electrical code [36], requires that cables and connectors for levels 2 and 3 be de-energized unless connected to an EV for charging. Typically, there are various types of charging systems for EVs, and generally, they are categorized as level 1, level 2 charging, and level 3 charging, as shown in Figure 2. All commercial EVs have the capability to charge using level 1 or level 2 charging systems. Level 1 and 2 charging are ranged from 2 to 20 kW single or three-phase to supply AC charging current of up to 80 A. Level 2 chargers are equipped with SAE J1772 AC charging port to charge the EV batteries. Level 3 charging, delivers a rapid DC charging method to charge the vehicle batteries at installed stations. This charging technique practices a 3-phase source with a DC output to the vehicle.

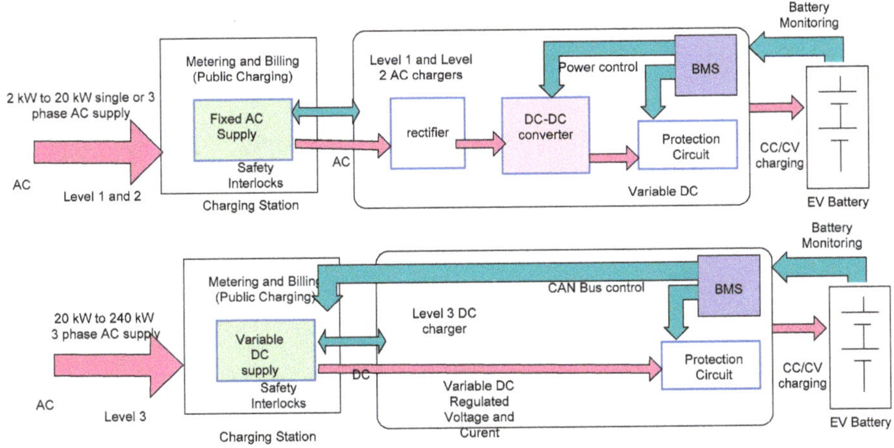

Figure 2. Block diagram of different types of charging systems.

2.2. Capacity Sizing of the Charging Station

In this paper, the charging station is considered far away from a community. The charging point in the station provides four typical DC chargers and two fast DC chargers (the fast charger typically requires 20 min to complete charging 80%, while the standard charger requires about 11 h to complete 80% charging). For example, the Nissan leaf car has about 24 kWh lithium-ion battery banks to store and supply power for EV motor [37–40]. Hence, within 30 min it reaches about 80% of its capacity at level 3 charging condition.

In this paper, the maximum capacity of the charging station output power was assumed. The load demand is treated as a changeable parameter and pursues normal allocation. For a public charging station, the chargers should endorse standard chargers, in order to be utilized for various types of EVs. Home charging usually uses AC charging and it includes two types of charging; level 1 charging (120 V) and level 2 charging (240). On the other hand, DC faster chargers are largely used in the commercial chargers stations. The capital cost of the charging station will be cut-price as the prices of wind and solar PV generator apparatus come down. Any overabundant energy from WT and PV can be accumulated in the energy storage batteries. The optimal sizing of WT, PV and battery capacities could depend on the variance of wind velocities and solar irradiance. Table 1 shows the renewable sources-based specifications of the charging station.

Table 1. The output of the charging station.

	Output	Qty	Working Hours
Faster DC charger	50 kW	2	24
DC standard charger	10 kW	4	24
Lights and other loads	10 kW		12
The total output	150 kW		

The charging station is designed, such as the maximal output power is 150 kW. The daily operation time is 24 h and seven days a week. The typical battery bank capacity is 24 kWh (Nissan Leaf, 2014). The calculated maximal demand for the charging station is 4728 kWh every day. The most popular charging technology, based on Japan's EV association standard [36,37], can deliver 50 kW output. However, the power rating of the charger is 90 kW at Tesla's supercharger station. Then the car can travel about 240 km after charging for 30 min. In this paper, it is assumed that the DC fast charger rates 50 kW output.

3. Control Strategy for PV/Wind/Storage Hybrid System

3.1. Control Scheme of the Boost DC-DC Converter Interfacing PV Array

The boost conversion stage is used to regulate the voltage from the PV panel and extract the maximum power. The PV panel voltage V_{pv} and the input current I_{pv} are sensed frequently. Then the MPPT control algorithm utilizes these two values and calculates the reference power that the PV panel requires to be operated at MPP conditions. The MPPT is achieved using an inner current loop and an outer voltage loop, as shown in Figure 3. By increasing the current drawn from the boost converter, results in reducing the panel output voltage. Therefore, the outer voltage compared with a reference value and feedback is regulated using PI controller gains. Hence, the output voltage is prevented from exceeding the adjusted value. On the other hand, the resulting signal from the MPPT controller is regulated using a PI controller. Then the output of the internal loop is compared with the reference current produced by the outer loop to generate the PWM signal [38–40].

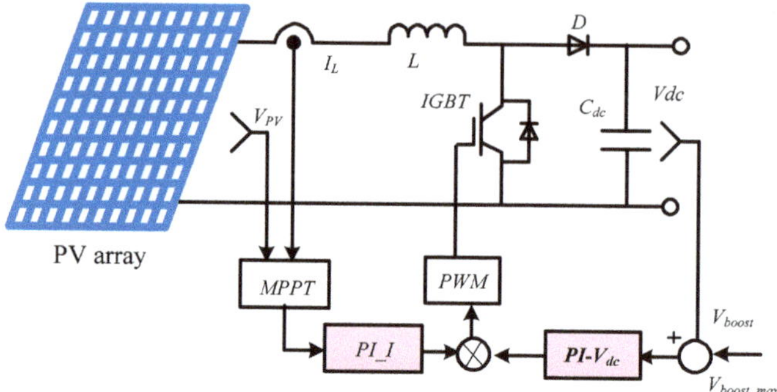

Figure 3. Maximum power point tracker (MPPT)control of boost DC-DC converter.

3.2. Control of the Boost DC-DC Converter Interfacing Wind Turbine

The control scheme of the WT generator includes maximum power point extractor for standalone variable speed WT with a permanent magnet synchronous generator (PMSG) and DC bus voltage control. The boost power converter is correctly adjusted to supply the maximum available generated power from the WT using the rectified DC voltage and current drawn from the rectifier output. The charging station works in standalone operation mode. Hence, the generated energy should be transferred through the DC microgrid to the electric vehicle loads. The power reference is generated from the comparison of the DC-link actual and reference values. The generated control signals are adjusted by using PI controllers to give power reference. For maintaining the DC microgrid voltage at its desired value, the PWM modulation signals of each converter are controlled regardless of variations in wind speeds and vehicles charging loads. The aim of using the boost converter is to regulate the rectified DC voltage to a higher voltage level for supplying generated power to the station DC microgrid. The DC microgrid voltage will be in the range of 280–320 V. DC-DC converters are controlled to obtain maximum power point operation MPP to maximize gathered wind power and to optimize the electrical energy produced by the PV panels. Figure 4 explains the control scheme of the boost converter. The parameters of PMSG are listed in Table 2. Thus, the measured input current and voltage values are used in the power optimizing algorithm or power tracker MPPT. The rectified DC voltage value (V_{DC}) is provided to a look-up table that defines a predefined maximum power point (MPP) characteristic curve.

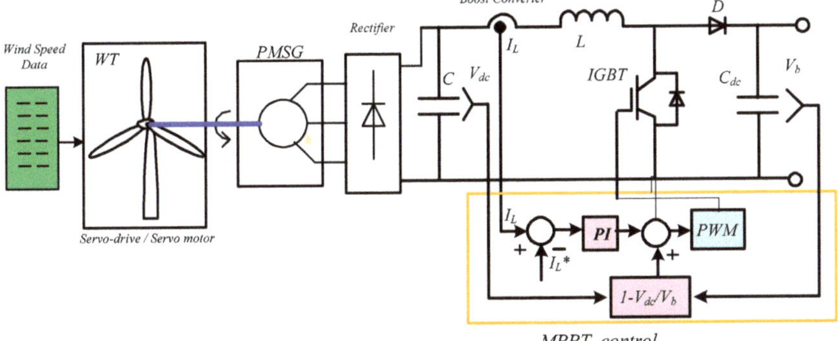

Figure 4. Control of the wind turbine (WT) boost converter interfacing DC microgrid.

Table 2. PMSG parameters [29].

Parameter	Value	Unit
Stator resistance	0.02	Ω
d-axis inductance *Ld*	7	mH
q-axis inductance	7	mH
V_{pk}/krpm	98.7	
No. of poles (P)	8	
Moment of inertia	8×10^{-3}	N-msec^2
Mechanical time constant	0.04	

3.3. Control of Bidirectional DC-DC Converter Interfacing Battery Bank

As shown in Figure 5, the bidirectional converter consists of a high-frequency inductor L, filtering capacitor C_{DC} and two half-bridge switches (S_1 and S_2), which enable a bidirectional flow of current. There are two voltage controllers with appropriate control blocks to realize the desired energy flow in various conditions. The controller produces a reference current of energy charging and discharging. The first controller is for DC-bus voltage regulation, and the second controller is for battery voltage control. To improve energy management in the charging station and the DC microgrid, backup energy storage batteries are used. The battery bank is connected to the DC-microgrid employing a bidirectional DC-DC power converter. This converter carries out double tasks: A battery charging regulator and a boost converter to supply power from the battery bank to the DC microgrid when the PV panels and wind sources have insufficient power to charge the electric vehicle loads. As a standalone charging station, the most convenient operating condition takes place when the electric vehicle power and the PV and wind extracted power agree. However, too deep discharge of the battery bank is not recommended, as, at a low battery bank voltage, there is a confined range of charging energy, which may cause over-voltage in DC microgrid during, e.g., energy recovery from the EVs side. On the other hand, there is a limited range of discharging energy, and the batteries have to be protected.

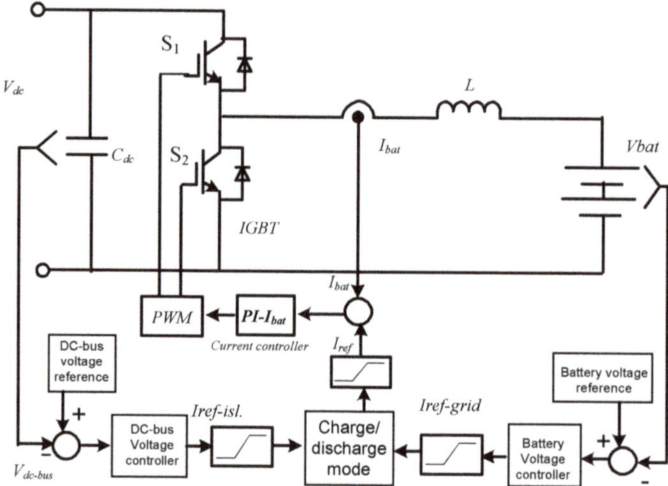

Figure 5. Control scheme of battery energy storage.

3.4. DC Microgrid Control Method

In the studied DC microgrid, a control scheme has been implemented to balance the DC voltage bus and to control the power supply to meet the load demand in islanded mode. In this control, one unit source acts as a master controlling the full system, while the rest of the units work as current sources (i.e., as "Slaves"). In this way, there will not be the voltage difference between the outputs of the DC sources, because the Master unit regulates the voltage values of all the output units; therefore, current will not circulate between the sources.

The DC microgrid is measured and compared with a predefined reference voltage, and the voltage error is processed through a compensator (PI block) to obtain the desired impedance current reference for the current loop. This compensator can be expressed in the following way [33]:

$$I_{Lref} = k_p(V_{ref} - V_{MG}) + k_i \int (V_{ref} - V_{MG})dt \tag{1}$$

where I_{Lref} is the reference current for the DC-DC converter. V_{ref} is the reference voltage for DC microgrid, and V_{MG} is the actual voltage. K_p and k_i are the proportional and integral controller coefficients for voltage loop. The power flow is controlled by a current controller who compares the impedance current in the master unit with the reference current desired to stabilize the system, the error is processed through another PI block to obtain the desired duty cycle for the converter which acts as a Master. The PI block can be expressed as:

$$d = k_{ip}(I_{Lref} - I_L) + k_{ii} \int (I_{Lref} - I_L)dt \tag{2}$$

where I_{Lref} is the reference current for the DC-DC converter. I_L is the actual measured current. K_{ip} and k_{ii} are the proportional and integral controller coefficients for the current loop. The problem of this control topology is the dependence on the master unit, and if there is a fault in this unit, the control will stop working properly [27]. To increase the reliability of the system, three different sources can act as a master unit, decreasing the chance to fault in the microgrid control. The energy storage system (ESS) can control the voltage level and the power flow through a bidirectional converter. When the microgrid is working in an islanded mode, this source will act as a "master" remaining the voltage at 300 V and meeting the load demand. If there is a fault in the ESS or the state of charge (SOC) level is not properly to control the microgrid in an islanded mode. There is a voltage controller implemented with a voltage and a current loop as it is shown in Figure 6.

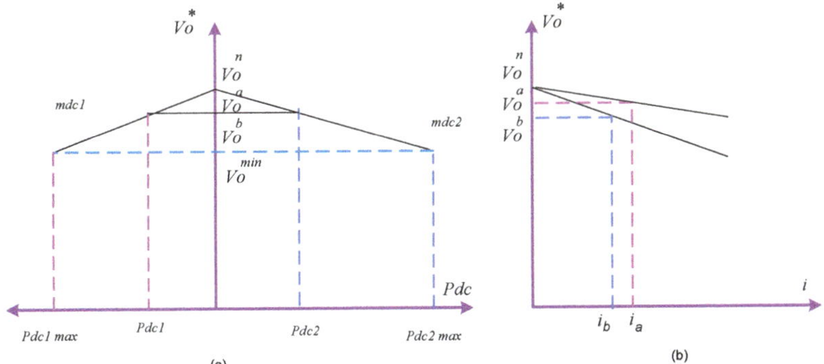

Figure 6. Representation of DC droop (**a**) The particular droop of two DG units; (**b**) v-I droop for the DC microgrid.

The DG units are interfaced through DC-DC converters to the DC microgrid. The individual droop-based power-sharing in the DC microgrid is represented in Figure 6a. The DC-link voltage of each DC-DC power converter is drooped with the DC generated power (P_{DCJ}) utilizing the droop coefficient m_{DCj} as represented in the equation:

$$V^*_{oj} = V^n_{oj} - m_{DCj}P_{DCj} \quad (3)$$

where V^*_{oj}, V^n_{oj} are the reference and no-load DC-link voltages of the DC-DC power converter, whereas the subscript j refers to a DG unit in the DC microgrid. The delivered DC power from each DG unit (Pacj) is wirelessly specified to supply the connected DC-load (P_{DCL}), following the equality

$$m_{DC1}P_{DC1} = m_{DC2}P_{DC2} \quad (4)$$

where $P_{DCL} = P_{DC1} + P_{DC2}$.

The primary requirement for a DC microgrid operation is to maintain the common DC-link voltage within a predefined range. Different measures shall be taken by each terminal of DC microgrid according to microgrid operation conditions. Therefore, a reliable and fast control scheme is essential for acknowledging system operation status. The DC-link voltage is a proper indicator of the DC microgrid's operational condition. An equivalent circuit of the DC-mircogrid, including the BESS and PEV is simplified, as shown in Figure 7, where P_{DC} and P_{AC} represent to the total power flow on the DC side of microgrid (PV panel, Battery bank and DC/DC power converters) and the AC-side (inverter and the AC load). From Figure 7, the instantaneous power relationship in the DC-microgrid is described by

$$P_{dc}(t) = P_{BESS}(t) + p_c(t) + P_{PEV}(t) + P_{ac}(t) \quad (5)$$

where P_{DC} is the DC power delivered by the DC-DC converter to the DC-microgrid, P_{BESS} is the power supplied to (or by) the BESS, Pc is the power to the DC-link capacitor, P_{PEV} is the power required for charging the plug-in electric vehicles PEV, and P_{AC} is the power required by the inverter for supplying the AC load.

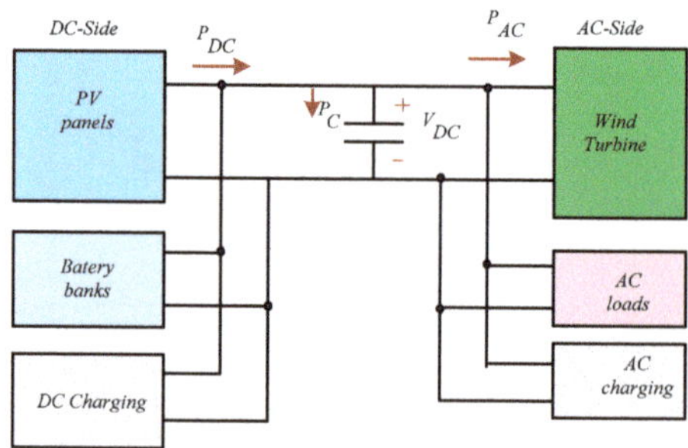

Figure 7. DC power flow diagram.

4. Energy Management Control Strategy

The amount of energy that can be produced by the PV or wind generator is calculated using the input data, such as hourly solar irradiance, wind speed, and ambient temperature. The total value of generated output power (Psources) is compared with the load demand energy (PPEVs) to estimate the energy flow distribution between the energy storage unit and the loads. Surplus energy is stored in the battery banks. The control strategy of the hybrid PV-wind charging station is described by the flowchart in Figure 8. According to Figure 8, the control strategy is applied according to four different cases as follows:

If P_{sourcs} $(P_{pv} + P_w) > P_{PEVs}$, then $P_{ESU} = P_{sources} - P_{PEVs}$. If the irradiance level and wind speed are high enough, the output power empowers the connected electric vehicles, and the exceeding power is stored in the battery bank.

If $P_{sourcs}{\cdot}(P_{pv} + P_w) = P_{PEVs}$, then $P_{ESU} = 0$. That is, if the irradiance level and wind speed are just enough, to empowers the connected electric vehicles and no excess power to charge the battery bank.

If $P_{sources} < P_{PEVs}$ and $P_{PEVs} - P_{sources} \leq P_{ESU}$, then $P_{PEVs} - P_{sources} = P_{ESU}$. That is, if the PV and wind generators cannot supply the load, then the load is supplied directly from the DC-microgrid, and the battery converter is switched on.

If $P_{sources} < P_{EVS}$ and $P_{EVS} - P_{sources} > P_{ESU}$, in this case, the energy stored in the battery bank is not enough to charge connected PEVs. Then the PEVS and battery bank are disconnected.

The supervisory controller is divided into two main functions. The first function identifies the mode of operation according to the conditions and situation of individual microgrid components. The second function is integrated into intelligent systems, such as converters and inverters that determine the performance of individual components in that mode of operation. The switch over in the battery charging mode takes place either when the state of charge of the battery is lower than the minimum SOC_{low} or when there is a sudden decrease in the required power for load and state of charge of the battery is lower than maximum SOC_{high}. Therefore, when loading power decreases, the surplus power is utilized to charge the batteries if it is not fully charged. The individual components can be controlled easily using a built-in controller, such as the DC-DC converter controller. Thus, the energy management system (EMS) is responsible for achieving the optimal operation of the DC microgrid.

The energy management control algorithms overcome the unbalance between power produced from distributed generation (DG) units and load. This can be done when the SOCs of ESS are sufficient. In the case of ESS failure or an inappropriate SOC value, the master unit becomes the wind turbine. In case of that, the load is greater than the available energy production, and the controller of the DC-microgrid is not able to balance the power flow of the system, the solution will be the load shedding. In case that the power generated by the sources is bigger than the load consumption, one of the distributed generators will be disconnected from the microgrid.

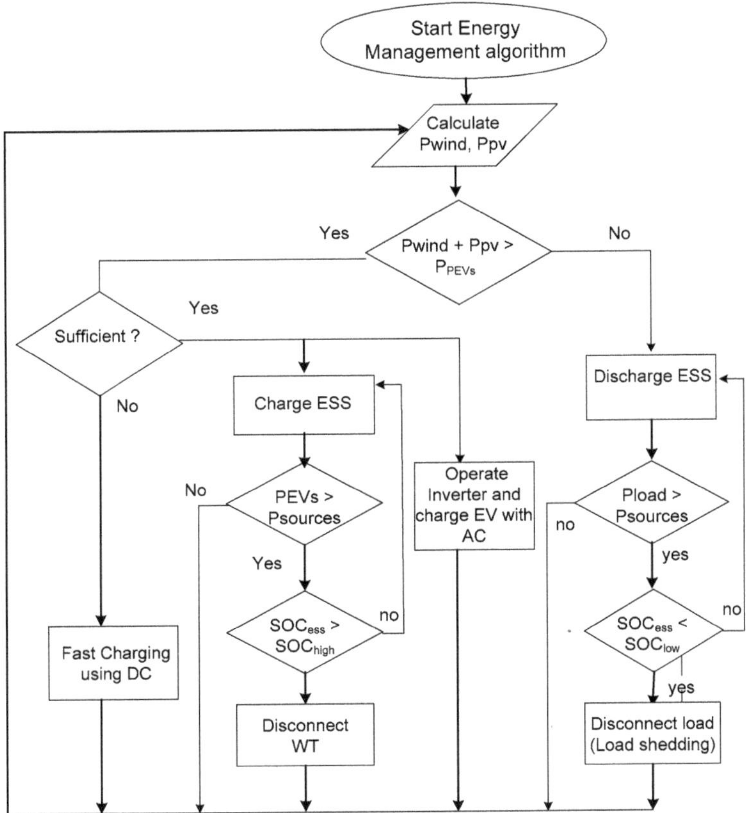

Figure 8. Flow chart of energy management control for the PV/wind/storage system.

The operation of the charging station DC-microgrid can be divided into four operation modes. The power flow direction is changing during various operation modes of the DC-microgrid based charging station, as illustrated in Figure 9. These modes of operation can be explained as follows:

Mode 1: $V_{DC} \geq V_{DC3}$: PEV charging and battery bank charging mode

The PV panels produce enough power, and this appears in an increase of the DC-link voltage to be higher than V_{DC3}. This additional energy produced by the PV panels and the wind turbine is supplied to the batteries through a bidirectional DC-DC converter. As soon as the PEVs are fully charged, all the power produced by the PV and wind sources is delivered to the battery banks.

Mode 2: $V_{DC1} > V_{DC} > V_{DC3}$: Charging by PV power

At this operating mode, the PEV is charged using the power generated by the PV system. In this case, the controller ensures that the PEV battery is not exceeding the over-charging limit. Thus, the controller terminates PEV charging when PEV voltage exceeds VBH (the voltage relating to 95 % state of charge of the PEV battery). This interval continues as long as the value of DC-link voltage is in a range between V_{DC1} and V_{DC3}.

Mode 3: $V_{DC3} \leq V_{DC} < V_{DC2}$: Wind turbine supplying power and battery bank discharging

During this mode, the power produced by the wind turbine is less than the required power for charging the PEV. Therefore, the whole power produced by the wind turbine is transferred to the PEV, and the additional amount is supplied by the battery bank. However, the DC-link voltage changes with the variation in solar irradiation and wind speed. Thus, any variation in the DC-link voltage at the DC-microgrid is monitored by the controller to produce a proper voltage at the output of the bidirectional DC-DC power converter. The renewable energy sources continue charging the PEV; whereas, the battery banks cover the peak load demand.

Mode 4: Case-1: $V_{DC} < V_{DC\text{-}1}$ and $I_{DMD} < I_{DMD\text{-max}}$

In this mode, the PV panels and wind turbines do not produce any power, due to inconvenient weather conditions. The boost DC-DC power converter is isolated, and the battery bank supplies the power required for charging PEVs. At any instant, during this mode, if the DC-link voltage VDC exceeds $V_{DC\text{-}1}$, the controller moves the system to work in Mode 2. The bidirectional DC-DC power converter regulates the output current and voltage for charging the PEV battery. As the battery bank is at off-peak, it continues to supply energy until the vehicles are completely charged. The controller terminates the charging process of PEV by disabling the DC-DC converter when the battery voltage V_{Bat} exceeds its maximum value V_{BH}.

Case-2: $V_{DC} < V_{DC\text{-}1}$ and $I_{DMD} \geq I_{DMD\text{-max}}$

This case is similar to case 1, but local demand exceeds the maximum demand of the microgrid. During this period, the PEV can be charged using the stored energy in the BESS if it is enough to cater to the charging process of PEVs. This continues until the state of charge of BESS (battery bank) decreases below its minimum value (SOC < SOCmin). At this moment, the charging process of PEVs is stopped tentatively by de-activating the bidirectional DC-DC power converter. Once the renewable energy power is back to off-peak conditions (i.e., $I_{DMD} < I_{DMD\text{-max}}$) the charging process of the PEVs is restored, and the controller supervises charging parameters.

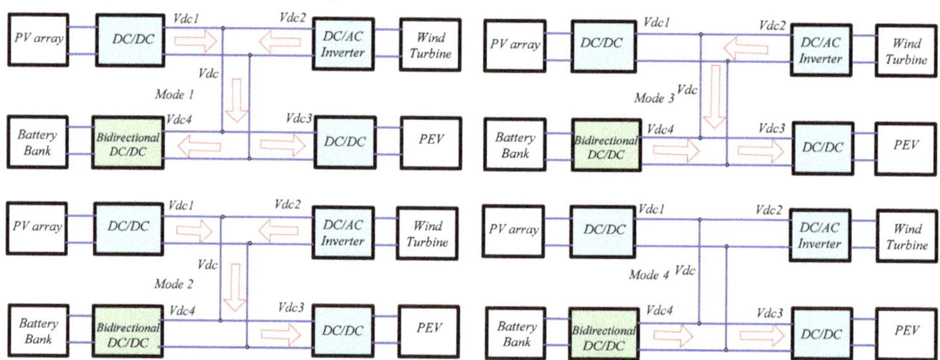

Figure 9. Different modes of operation.

An example of the electric vehicle load supplied by the DC-microgrid is shown in Figure 10 for one day period. Figure 11 shows the solar irradiance, temperature and wind velocity profiles during a typical day in Sohag city, Egypt. The wind speed profile starts with 5 m/s at time 1 h, rises to 9 m/s at time 6 h, then falls down to 5 m/s at time 10 h, etc., as shown in Figure 11. The generators meet the load during the night. In the morning when the sun comes up, and the PV starts generating, the PV charges the BESS until the PV generation and the BESS state of charge are high enough that they can meet the load on their own without the generators. The load is then transferred to the PV/BESS system, and the generators turn off. The BESS and PV power the load together from 7–8 a.m. when PV generation is not yet high enough to meet the load by itself. At 8 a.m., when PV can fully meet the load, the BESS stops discharging. Excess PV generation is used first to charge the BESS and then remaining excess is curtailed. In the evening, PV generation decreases, until the PV and BESS can no longer meet the full load. Some of the PV generation is curtailed because the BESS is already fully charged. At this point, the generators turn on again and supply the load overnight.

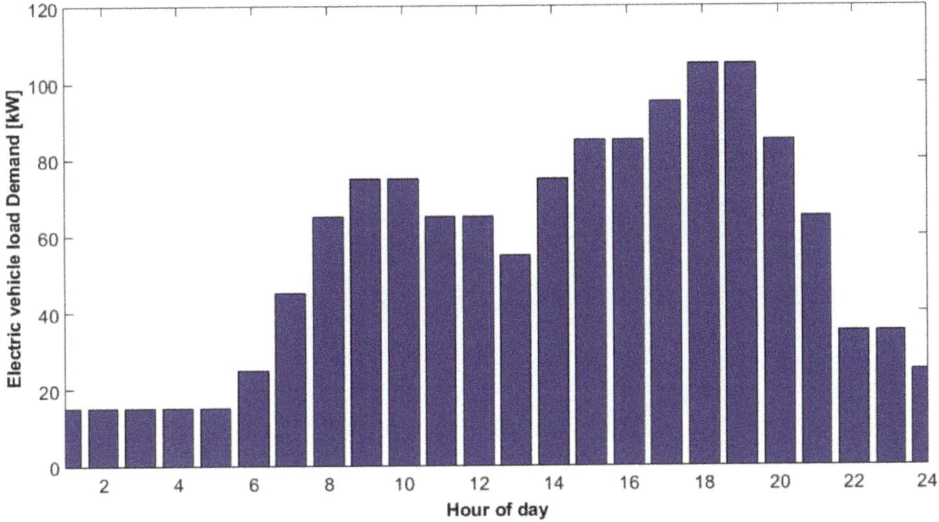

Figure 10. Electric vehicle load profile supplied by the DC-microgrid.

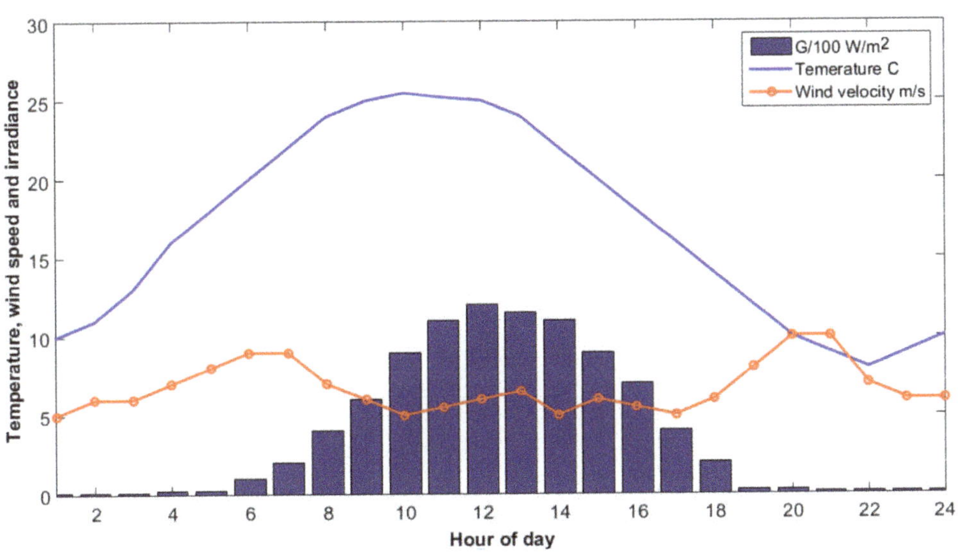

Figure 11. The solar radiation, temperature and wind velocity profiles during a typical day.

5. Simulation Results and Discussion

Simulation results are obtained based on the typical daily load profile of the studied EV presented in Figure 11. The calculated produced renewable power and load during a typical day in the studied system is shown in Figure 12. The hybrid system model is verified by implementing the detailed models in a MATLAB/ Simulink environment. This model presents an alternative emergency power system based on lithium-ion batteries. This model also features an energy management system for hybrid electric sources. The energy management system regulates the power between the energy sources and loads according to a predetermined control strategy. The Simulink model of the studied DC microgrid is shown in Figure 13. The specifications of the studied DC-microgrid are shown in Table 3:

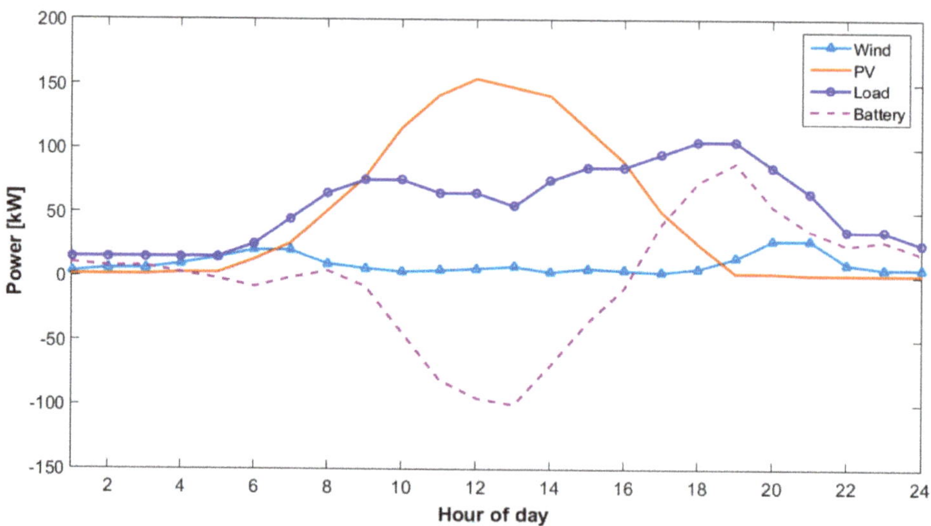

Figure 12. Daily profiles of renewable power generation and load in the studied system.

Table 3. System parameters.

Item	Description
PV Array	composed of 330 modules SunPower SPR-305E-WHT-D with series and parallel combination (Nser = 5 Npar = 66) rating 100 kW
Wind turbine	Rated output power = 10 kW Wind speed base = 12 m/s Base rotational speed = 500 rpm Initial rotational speed = 200 rpm Moment of inertia = 0.08 p.u
Li-ion battery	A 48 V, 500 Ah, system
Battery state of charge	SOCmin–SOCmax: 60–90 [%]
Bidirectional DC-DC converter	A 50 kW, A controlled voltage/current outputs
Inverter system	A 150 kVA, 270 V DC in, 200 V AC, 60 Hz

The DC microgrid voltage is shown in Figure 14. The DC-link voltage is an indication for the DC generated power. Figure 15 represents the output voltage and output current of PV panels. The photovoltaic power generation is set to the Maximal Power Point Tracking, which is proportional to the irradiance solar radiation and (W/m^2). These typical weather data at intervals of one hour are collected. There is much meteorological software can estimate the solar radiation and the ambient temperature. In this figure, the solar irradiation is reduced from 1000 W/m^2 to 850 W/m^2 at time 2 s. Consequently, the total generated current from PV panels is reduced from 29.6 A to 25 A. The battery bank compensates the fluctuations of the difference between the microgrid reference power and all the passive power variations of the DC microgrid (PV/wind power and total loads). The battery bank voltage, charging and discharging current is shown in Figure 16. The corresponding state of charge SOC of the battery bank is shown in Figure 17.

The wind turbine model comprises mathematical models of wind turbines and wind speed simulation. Figure 18 shows how the voltage at generator terminals (instantaneous value) changes with time. Figure 18 shows the corresponding generated current. The output power of the wind generator is proportional to the cube of the wind speed. A sudden variation of wind speed from 12 m/s to 9 m/s happens at a time of 3 s. However, the temporal variations of the PMSG rotational speed, torque, voltage, and output power follow that of the wind speed. The rectified output voltage of the wind generator is shown in Figure 19.

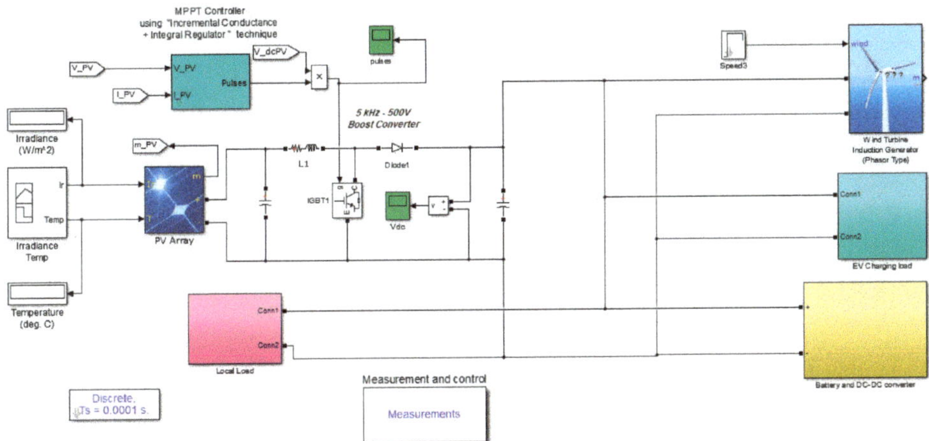

Figure 13. The Simulink model of the studied DC microgrid.

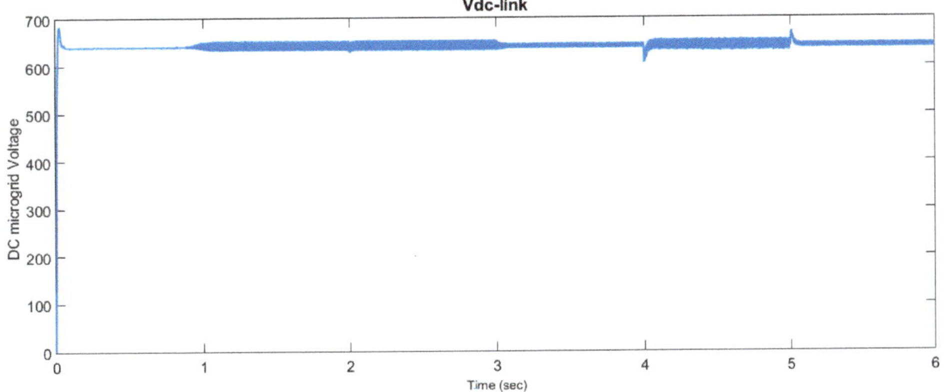

Figure 14. DC-microgrid DC-link voltage.

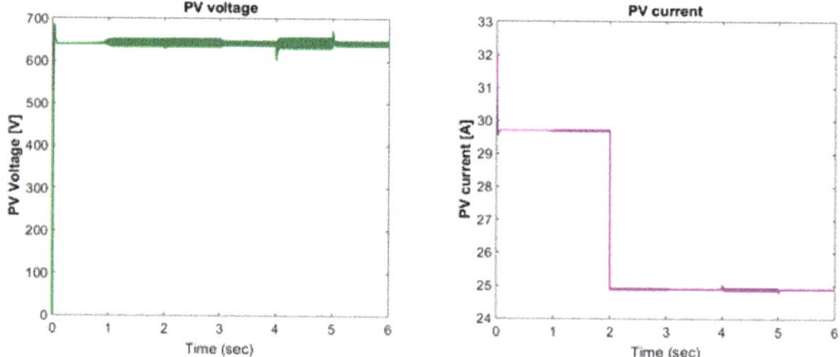

Figure 15. PV panels output voltage and current.

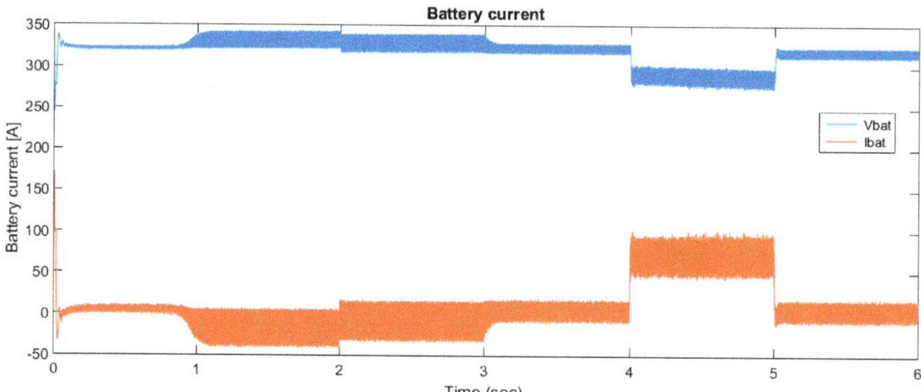

Figure 16. Battery bank voltage and current.

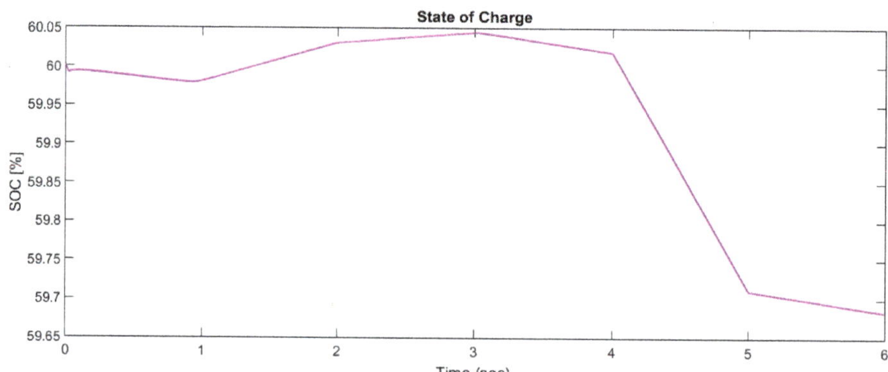

Figure 17. Battery bank SOC.

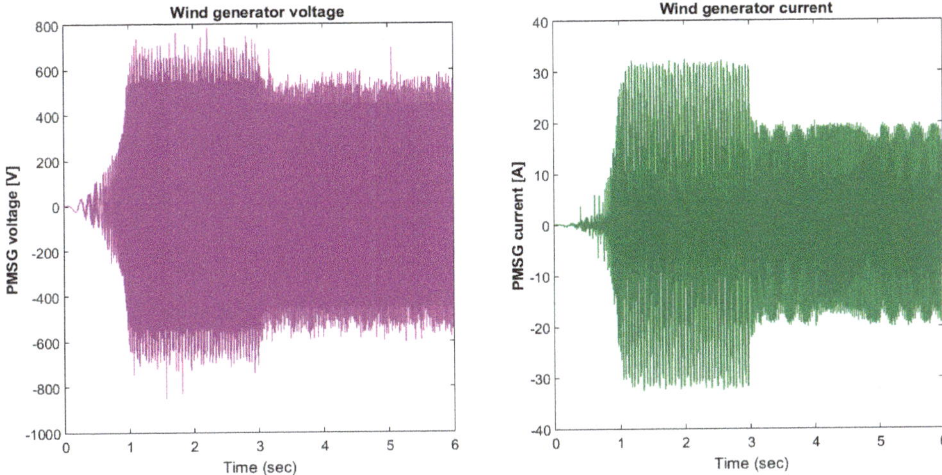

Figure 18. Wind generator output voltage and current.

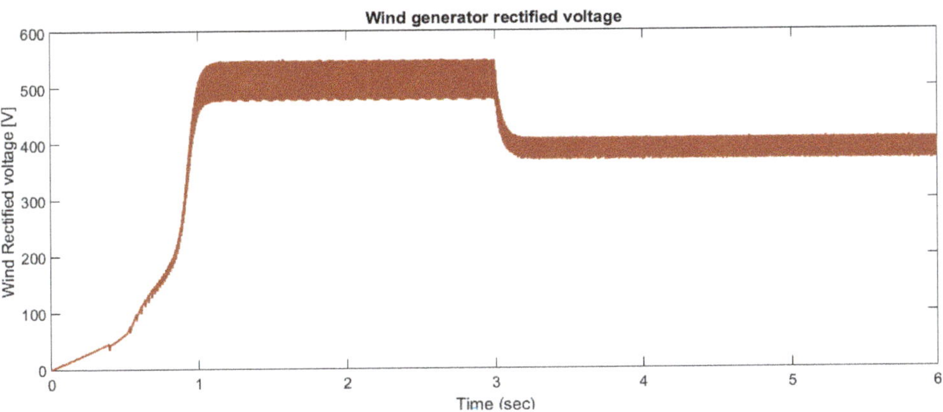

Figure 19. The rectified voltage of PMSG output.

The voltage in the line and the current fed to the inverter for AC charging are shown in Figures 20 and 21, respectively. The current drawn from the AC system is sinusoidal, due to the AC filters employed. The lack of such AC filters will directly feed the harmonics into the grid source. A 3-phase AC load is used to emulate the EV charging load profile. The load profiles were generated using SIMULINK, and then the hourly energy results were configured into a suitable format. For each month, three day types were used to represent the annual load: Peak day, weekday, and weekend. The total powers from different sources and loads are shown in Figure 22. A step change in load power occurs at time 4–5 s from 20 kW to 60 kW. During this period, the peak power is compensated from the battery bank. The electrical power performance, current-voltage characteristics and system response confirm that the system has satisfactory performance under conditions of a step changing power reference and loads disturbances.

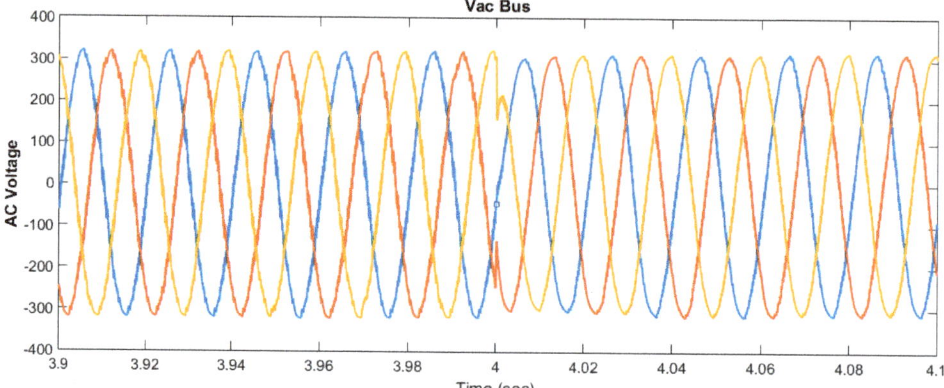

Figure 20. AC side voltage for AC charging.

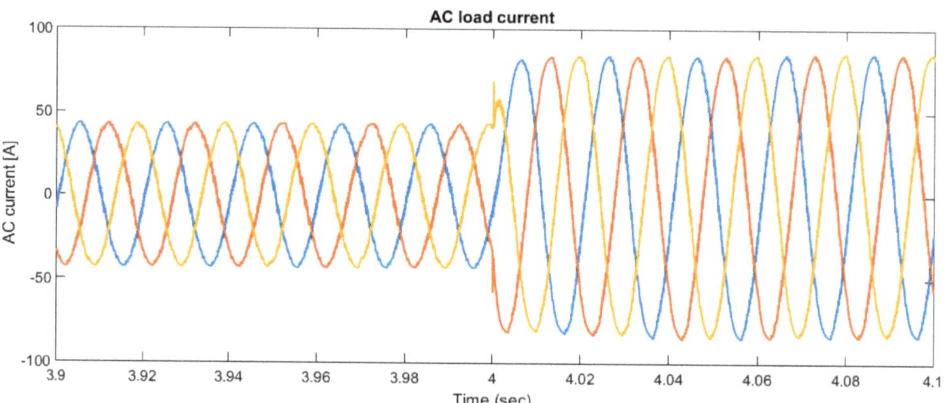

Figure 21. AC side charging current.

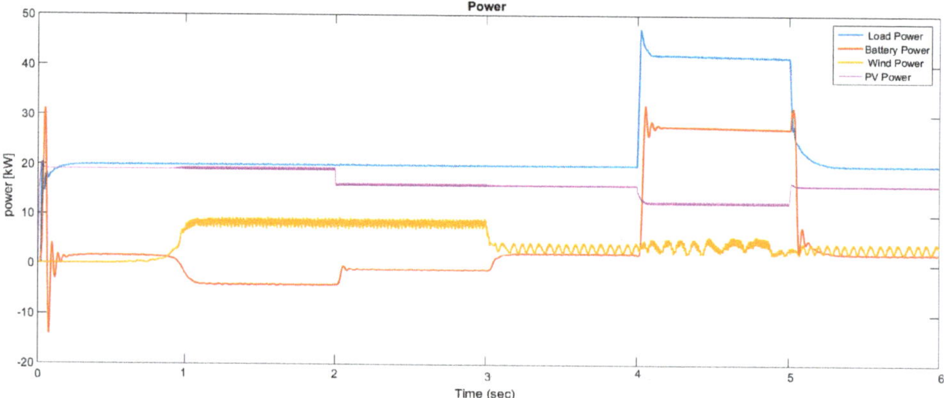

Figure 22. Different power values for sources and loads.

6. Resilient DC-Microgrid

DC microgrid configurations are evaluated that can be used to integrate PV and wind generators alongside existing batteries' energy storage systems (BESS) to increase resiliency at the site. The BESS units are sized to support the charging station load for one hour in this operation condition. There will be cloudy intervals or early morning/ late afternoon hours when PV and wind generators will not be capable of delivering the required charging load. During those times, the BESS is required to supply the EV loads until PV or wind generation supplies the charging station load. An integrated solution was proposed so that all energy sources operate in an integrated manner and are centrally controlled. Therefore, the BESS does not require to be sized for the full load in this operation scenario. The DC-bus Voltage of microgrid effects time of charging. Then the time of charging can be calculated according to the voltage and capacity of the battery. This measure takes into account the robustness of the system against disturbances and the quickness of the recovery [27].

$$R = \frac{\int_{t_0}^{T_a} V_{dc-bus}(t)dt}{T_a} \tag{6}$$

where $V_{DC\text{-}bus}$ is the DC bus voltage. R is the resiliency, T_a is the recovery time.

The resilience of an electrical microgrid can be defined as "the ability of the microgrid to sustain against disturbances and return to its normal state quickly". This definition includes the two remarkable attributes, recovery and response, and is compatible with the definitions given in references [27,28]. The schematic diagram of the system's resilience concept is illustrated in Figure 10. However, disruption occurs at time t_0, as shown in this figure, and the system performance (DC bus voltage) falls from Q_0 to Q_1. By taking adequate action, the system returns finally to original DC-link voltage at time t_1.

As shown in Equation (6), the resilience measure is able to comprehensively represent the ability of the system to withstand the disruption and recover rapidly. Here, $0 < R < 1$. Therefore, when $R = 1$, it means that the system has perfect resilience: Either its performance degradation is 0, or it can recover from disruption instantaneously. In case $R = 0$, it designates that the system is completely troubled immediately upon disruption and cannot recover within the maximum permissible recovery time. It is obvious that systems with higher values of R are more resilient.

The performance curve $Q(t)$ is used to describe the system resilience of microgrid. The performance loss function from disruption is defined by the integral of the curve, followed by a gradual recovery (i.e., the shadowed area in Figure 23). This measure achieves the robustness of the system versus disturbances (load disturbance and intermitted generated energy from renewable sources) and the quickness of the recovery action. By calculating the area under the curve of Figure 24, the DC-bus voltage is recovered from V_{DC} = 96% to 100% and the time from 4 to 4.06 s and dividing this by time 0.06; then the resiliency will be 0.98. It means the system is near perfect resilient.

A new resilience measure is proposed in this paper for DC microgrid. It comprises using the maximum admissible recovery time as the considered time interval and enabling an estimation method. Resilience measurement scheme is used to estimate the resilience of different microgrid designs. It is also used to verify whether the resilience goal of microgrid can be satisfied, and choose a resilient method that can sustain the disruption and return the microgrid to the normal state as quickly as possible.

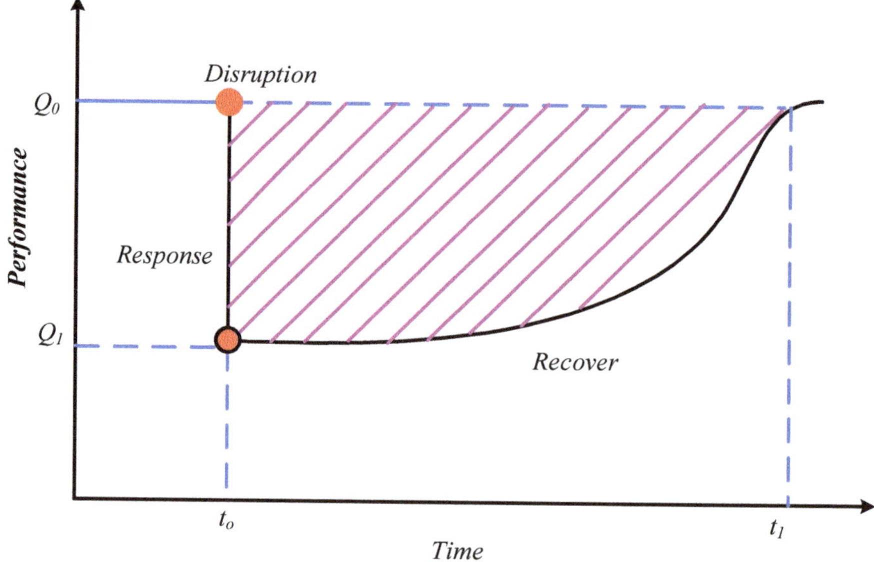

Figure 23. The schematic representation of resilience [27].

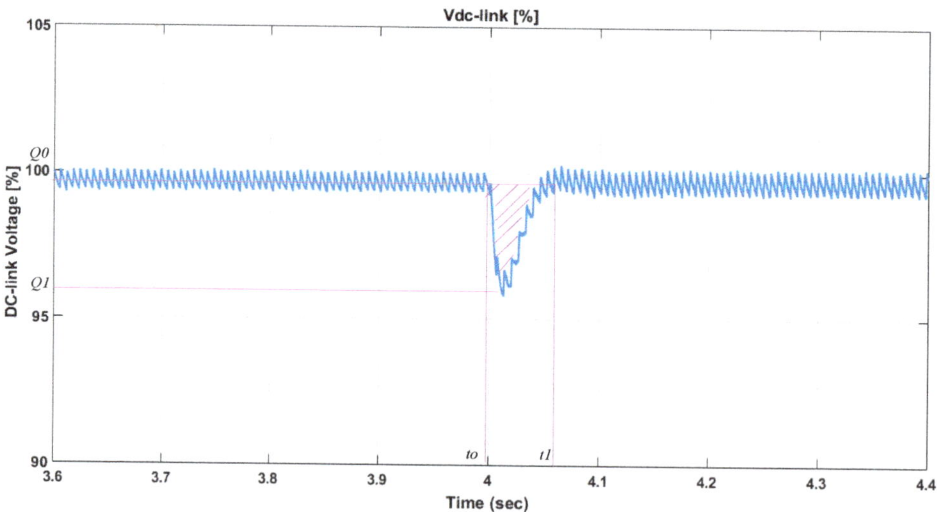

Figure 24. Dc-link voltage for calculating the resilience.

7. Conclusions

This paper has presented an energy management technique for an isolated DC-microgrid supplying EV charging station. The standalone DC microgrid is verified by implementing the charging station model in a MATLAB/Simulink environment. The EMS control system is designed to regulate renewable energy sources status, battery SOC, and load demand. For an accurate evaluation of EMS strategy, hourly variations of renewable generation and a typical EV load are utilized as input data. The solar irradiation is reduced from 1000 W/m^2 to 850 W/m^2 at time 2 s. A sudden variation of wind speed from 12 m/s to 9 m/s happened at a time of 3 s. However, the temporal variations of the PMSG rotational speed, torque, voltage, and output power follow that of the wind speed. A step change is applied in the load power then the battery bank compensates the fluctuations of the difference between the microgrid reference power and all power variations of the DC microgrid (PV/wind power and total loads). The electrical power performance, current-voltage characteristics and system response confirm that the system has satisfactory performance under conditions of a step-changing power reference and loads disturbances. The results indicated that the integration of intermittent wind and solar energy sources in microgrid should be designed carefully in standalone operation. The proposed control strategies can provide excellent performance under different operating conditions. BESSs can increase the reliability of the system because they can store excess renewable energy during low-demand periods and can supply during high-demand periods.

Author Contributions: Conceptualization, K.S.; Methodology, A.G.A.-K. and A.S.A.; Software, A.G.A.-K.; Validation, K.S.; Formal Analysis, A.S.A.; Investigation, A.G.A.-K.; Resources, A.S.A.; Data Curation, K.S.; Writing-Original Draft Preparation, A.G.A.-K.; Writing—Review and Editing, A.S.A.; Visualization, K.S.; Supervision; Project Administration, A.G.A.-K.; Funding Acquisition, A.S.A.

Funding: This research received no external funding.

Acknowledgments: The authors extend their appreciation to the Deanship of Scientific Research at Majmaah University for funding this work under project number No (RGP-2019-19).

Conflicts of Interest: The authors declare no conflict of interest.

Abbreviations

The following abbreviations are used in this manuscript:

DC	Direct Current
AC	Alternating Current
EMI	Electromagnetic Interference
DG	Distributed Generation
BMS	Battery Management Systems
SOC	State of charge
EV	Electric vehicle
PMSG	Permanent Magnet Synchronous Generator
IDMD	Maximum demand current
PEV	Plug in Electric Vehicle
BESS	batteries energy storage system
EMS	Energy management Strategy
EMU	Energy management unit
ESS	Energy Storage system
DG	distributed generation
MPP	Maximum power point
MPPT	Maximum power point tracking
PV	Photovoltaic
PI	Proportional-integral
WT	Wind Turbine
WTCS	Wind Turbine Conversion System
DERs	Distributed Energy Resources
HVAC	heating, ventilation, and air conditioning
PWM	pulse-width modulation
RE	Renewable Energy

References

1. Sayed, K.; Gabbar, H.A. Electric Vehicle to Power Grid Integration Using Three-Phase Three-Level AC/DC Converter and PI-Fuzzy Controller. *Energies* **2016**, *9*, 532. [CrossRef]
2. Sayed, K. Zero-voltage soft-switching DC–DC converter-based charger for LV battery in hybrid electric vehicles. *IET Power Electron.* **2019**, *12*, 3389–3396. [CrossRef]

3. Long, B.; Lim, S.T.; Bai, Z.F.; Ryu, J.H.; Chong, K.T. Energy management and control of electric vehicles, using hybrid power source in regenerative braking operation. *Energies* **2014**, *7*, 4300–4315. [CrossRef]
4. Fan, Y.; Zhu, W.; Xue, Z.; Zhang, L.; Zou, Z. A multi-function conversion technique for vehicle-to-grid applications. *Energies* **2015**, *8*, 7638–7653. [CrossRef]
5. Lukic, S.M.; Cao, J.; Bansal, R.C.; Fernando, R.; Emadi, A. Energy storage systems for automotive applications. *IEEE Trans. Ind. Electron.* **2008**, *55*, 2258–2267. [CrossRef]
6. EPRI. Tennessee Valley Authority Smart Modal Area Recharge Terminal (SMART) Station Project. 2012. Available online: http://www.epri.com/abstracts/Pages/ProductAbstract.aspx?ProductId=000000000001026583 (accessed on 25 December 2012).
7. Gorton, M. Solar-Powered Electric Vehicle Charging Stations Sprout up Nationally. 2011. Available online: http://www.renewableenergyworld.com/rea/news/article/2011/11/solar-powered-electric-vehiclecharging-stations-sprout-up-nationally. (accessed on 15 November 2011).
8. Ibrahim, H.; Sayed, K.; Kassem, A.; Mostafa, R. A new power management strategy for battery electric vehicles. *IET Electr. Syst. Transp.* **2018**, *9*, 65–74.
9. Savio, D.A.; Juliet, V.A.; Chokkalingam, B.; Padmanaban, S.; Holm-Nielsen, J.B.; Blaabjerg, F. Photovoltaic Integrated Hybrid Microgrid Structured Electric Vehicle Charging Station and Its Energy Management Approach. *Energies* **2019**, *12*, 168. [CrossRef]
10. Jha, M.; Blaabjerg, F.; Khan, M.A.; Kurukuru, V.S.B.; Haque, A. Intelligent Control of Converter for Electric Vehicles Charging Station. *Energies* **2019**, *12*, 2334. [CrossRef]
11. Liu, K.; Makaran, J. Design of a solar powered battery charger. In Proceedings of the Electrical Power and Energy Conference (EPEC), Montreal, QC, Canada, 22–23 October 2009.
12. Neumann, H.; Schar, D.; Baumgartner, F. The potential of photovoltaic carports to cover the energy demand of road passenger transport. *Prog. Photovolt. Res. Appl.* **2012**, *20*, 639–649. [CrossRef]
13. Ingersoll, J.G.; Perkins, C.A. The 2.1 kW photovoltaic electric vehicle charging station in the city of Santa Monica, California. In Proceedings of the Twenty Fifth IEEE Photovoltaic Specialist's Conference, Washington, DC, USA, 13–17 May 1996.
14. Locment, F.; Sechilariu, M.; Forgez, C. Electric vehicle charging system with PV grid-connected configuration. In Proceedings of the IEEE Vehicle Power and Propulsion Conference (VPPC), Lille, France, 1–3 September 2010.
15. Letendre, S. Solar electricity as a fuel for light vehicles. In Proceedings of the 2009 American Solar Energy Society Annual Conference, Boulder, CO, USA, 11–16 May 2009.
16. Tulpule, P.J.; Marano, V.; Yurkovich, S.; Rizzoni, G. Economic and environmental impacts of a PV powered workplace parking garage charging station. *J. Appl. Energy* **2013**, *108*, 323–332. [CrossRef]
17. Birnie, D.P., III. Solar-to-vehicle (S2 V) systems for powering commuters of the future. *J. Power Sources* **2009**, *186*, 539–542. [CrossRef]
18. Zhang, Q.; Tezuka, T.; Ishihara, K.N.; Mclellan, B.C. Integration of PV power into future low-carbon smart electricity systems with EV and HP in Kansai Area, Japan. *J. Renew. Energy* **2012**, *44*, 99–108. [CrossRef]
19. Ross, M.; Hidalgo, R.; Abbey, C.; Joos, G. Energy storage system scheduling for an isolated microgrid. *IET Trans. Renew. Power Gener.* **2011**, *5*, 117–123. [CrossRef]
20. Guo, L.; Liu, W.; Li, X.; Liu, Y.; Jiao, B.; Wang, W.; Wang, C.; Li, F. Energy Management System for Stand-Alone Wind-Powered-Desalination Microgrid. *IEEE Trans. Smart Grid* **2016**, *7*, 1079–1087. [CrossRef]
21. Kim, J.-Y.; Jeon, J.-H.; Kim, S.-K.; Cho, C.; Park, J.H.; Kim, H.-M.; Nam, K.-Y. Cooperative control strategy of energy storage system and microsources for stabilizing the microgrid during islanded operation. *IEEE Trans. Power Electron.* **2010**, *25*, 3037–3048.
22. Lu, X.; Guerrero, J.M.; Sun, K.; Vasquez, J.C. An improved droop control method for DC microgrids based on low bandwidth communication with dc bus voltage restoration and enhanced current sharing accuracy. *IEEE Trans. Power Electron.* **2014**, *29*, 1800–1812. [CrossRef]
23. Denholm, P.; Kuss, M.; Margolis, R.M. Co-benefits of large scale plug-in hybrid electric vehicle and solar PV deployment. *J. Power Sources* **2012**, *236*, 350–356. [CrossRef]
24. *Energy Resilience: Operations, Maintenance, & Testing (OM&T) Strategy and Implementation Guidance*; Office of the Assistant Secretary of Defense (Energy, Installations, & Environment): Washington, DC, USA, March 2017.
25. Helmus, J.; van den Hoed, R. Key Performance Indicators of Charging Infrastructure. In Proceedings of the EVS29 Symposium, Montréal, QC, Canada, 19–22 June 2016.

26. Bahramirad, S.; Khodaei, A.; Svachula, J.; Aguero, J.R. Building Resilient Integrated Grids. *IEEE Electrif. Mag.* **2015**, *3*, 48–55. [CrossRef]
27. Li, R.; Dong, Q.; Jin, C.; Kang, R. A New Resilience Measure for Supply Chain Networks. *Sustainability* **2017**, *9*, 144. [CrossRef]
28. Zobel, C.W. Representing perceived tradeoffs in defining disaster resilience. *Decis. Support Syst.* **2011**, *50*, 394–403. [CrossRef]
29. Yin, C.; Wu, H.; Sechilariu, M.; Locment, F. Power Management Strategy for an Autonomous DC Microgrid. *Appl. Sci.* **2018**, *8*, 2202. [CrossRef]
30. Al-Sakkaf, S.; Kassas, M.; Khalid, M.; Abido, M.A. An Energy Management System for Residential Autonomous DC Microgrid Using Optimized Fuzzy. *Energies* **2019**, *12*, 1457. [CrossRef]
31. Farzin, H.; Fotuhi-Firuzabad, M.; Moeini-Aghtaie, M. Enhancing Power System Resilience Through Hierarchical Outage Management in Multi-Microgrids. *IEEE Trans. Smart Grid* **2016**, *7*, 2869–2879. [CrossRef]
32. Che, L.; Zhang, X.; Shahidehpour, M. Resilience Enhancement with DC Microgrids. In Proceedings of the 2015 IEEE Power & Energy Society General Meeting, Denver, CO, USA, 26–30 July 2015. [CrossRef]
33. Sayed, K.; Gabbar, H. Supervisory Control of a Resilient DC Microgrid for Commercial Buildings. *Int. J. Process. Syst. Eng.* **2017**, *4*, 99–118. [CrossRef]
34. Sayed, K.; Abdel-Salam, M. Dynamic performance of wind turbine conversion system using PMSG-based wind simulator. *Electr. Eng. J.* **2017**, *99*, 431–439. [CrossRef]
35. Sayed, K.; Gabbar, H.; Nishida, K.; Nakaoka, M. A New Circuit Topology for Battery Charger from 200V DC Source to 12V for Hybrid Automotive Applications. In Proceedings of the 2016 IEEE Smart Energy Grid Engineering (SEGE), Oshawa, ON, Canada, 21–24 August 2016; pp. 317–321.
36. Nissan Leaf. Available online: https://en.wikipedia.org/wiki/Nissan_Leaf/ (accessed on 20 October 2018).
37. Japan Electric Vehicle Association Standards JEVS. Available online: http://www.evaap.org/ (accessed on 16 October 2018).
38. Abo-Khalil, A.G.; Alyami, S.; Sayed, K.; Alhejji, A. Dynamic Modeling of Wind Turbines Based on Estimated Wind Speed under Turbulent Conditions. *Energies* **2019**, *12*, 1907. [CrossRef]
39. Sayed, K.; Kwon, S.; Nishida, K.; Nakaoka, M. New DC Rail Side Soft-Switching PWM DC-DC Converter with Voltage Doubler Rectifier for PV Generation Interface. In Proceedings of the 2014 International Power Electronics Conference IPEC, Hiroshima, Japan, 18–21 May 2014; pp. 2359–2365.
40. Kumar, M.; Srivastava, S.C.; Singh, S.N. Control Strategies of a DC Microgrid for Grid Connected and Islanded Operations. *IEEE Trans. Smart Grid* **2015**, *6*, 1588–1601. [CrossRef]

Article

Energy Management of Hybrid Diesel/Battery Ships in Multidisciplinary Emission Policy Areas

Mohsen Banaei [1], Fatemeh Ghanami [1], Mehdi Rafiei [2], Jalil Boudjadar [2,*] and Mohammad-Hassan Khooban [2]

[1] Department of Electrical Engineering, Faculty of Engineering, Ferdowsi University of Mashhad, Mashhad 91775-1111, Iran; banaei.mohsen@gmail.com (M.B.); f.ghanami@mail.um.ac.ir (F.G.)

[2] Department of Engineering, Aarhus University, 8200 Aarhus, Denmark; Rafiei@eng.au.dk (M.R.); khooban@eng.au.dk (M.-H.K.)

* Correspondence: jalil@eng.au.dk

Received:11 July 2020; Accepted: 10 August 2020; Published: 12 August 2020

Abstract: All-electric ships, and especially the hybrid ones with diesel generators and batteries, have attracted the attention of maritime industry in the last years due to their less emission and higher efficiency. The variant emission policies in different sailing areas and the impact of physical and environmental phenomena on ships energy consumption are two interesting and serious concepts in the maritime issues. In this paper, an efficient energy management strategy is proposed for a hybrid vessel that can effectively consider the emission policies and apply the impacts of ship resistant, wind direction and sea state on the ships propulsion. In addition, the possibility and impact of charging and discharging the carried electrical vehicles' batteries by the ship is investigated. All mentioned matters are mathematically formulated and a general model of the system is extracted. The resulted model and real data are utilized for the proposed energy management strategy. A genetic algorithm is used in MATLAB software to obtain the optimal solution for a specific trip of the ship. Simulation results confirm the effectiveness of the proposed energy management method in economical and reliable operation of the ship considering the different emission control policies and weather condition impacts.

Keywords: hybrid diesel/battery ships; energy management; emission management

---

1. Introduction

Shipping has been a significant human action throughout history, especially in international and inter-regional trading applications. Different types of technologies have been used for providing the propulsion force of ships so far. All-electric ships are one of the recently introduced technologies which is referred to the ships that use electricity to provide the propulsion force [1]. Required electricity of these ships can be generated by different resources like diesel and gas power generators, batteries, and clean energies like fuel cell, or by hybrid of these resources [2–5]. At the moment, the hybrid of diesel generators and batteries is known as a popular method for supplying the all-electric ships loads and reducing the emission [6,7].

Energy management of all-electric ships has been thoroughly studied in the literature, so far. In [8], the authors use unconstrained, large-scale, global optimization to solve the energy resources scheduling problem of a large green ship with diesel power generator, batteries, photovoltaic panels and cold ironing as energy resources. A dynamic programming approach is used in [9] to solve the energy management problem of an all-electric ship with hybrid diesel/battery system considering emission limitations. A nonlinear procedure is used to achieve a control strategy of all-electric powered ships with only a hybrid energy storage system as the energy resource. In order to address shipboard load fluctuations, the authors of [10] apply a real-time model predictive control based on energy management strategy. They also use an integrated perturbation analysis and sequential quadratic programming algorithm to solve the optimal scheduling problem of a ship with hybrid batteries and ultra-capacitors. In [11], a combined cooling heat and power plant in the hybrid diesel/battery ships is introduced and a multi-energy configuration for it is proposed. In order to reduce the ship operating cost and gas emissions simultaneously, a multi-objective optimization problem is solved. In [12], a non-linear method is used to find the optimal control strategy of an all-electric ship supplied with a hybrid storage system. Shipboard loads power scheduling problem is solved using the particle swarm optimization method in [13]. Energy management of hybrid fuel cell/battery system for propulsion of ships has also appeared in the literature recently. In [14], a rule-base method is proposed for energy management of a hybrid fuel cell/battery ship. Simultaneous optimal component sizing and energy management of a hybrid fuel cell/battery ship are studied in [15]. Reference [16] applies a multi-scheme energy management method for optimizing the total energy consumption of a hybrid FC and battery system. Reference [17] proposes a non-linear model for optimal energy management of a hybrid fuel cell/battery ship under different operation scenarios.

Reviewing the above mentioned studies shows that there are still some gaps in the field of the energy management methods for all-electric ships. First of all, although it has a strong impact on the propulsion load, ship resistance due to sea waves is not discussed and formulated in detail in these studies. The impacts of wind direction and sea state of the operation of the ships are not formulated in the methods. Moreover, the ability of all-electric ships that carry vehicles in providing the possibility of charging to electric vehicles is not formulated in the aforementioned studies. This paper focuses on integrating impacts from water resistance and wind on the propulsion load. So, the main contributions of the paper are summarized as follows:

- An energy management method for the ships sailing in Inter-regional areas considering different emission constraints in different areas of the traveling route.
- Formulating the water resistant for the understudy ship in detail and integrating the resistance formulations in the energy management problem.
- Considering the impacts of wind direction and sea state on the efficiency of electric motors.

The rest of the paper is organized as follows. Section 2 presents the problem description. In Section 3, ship's resistance is formulated. The relation between fuel consumption and power is determined in Section 4. The problem is formulated in Section 5. Numerical results are presented in Section 4 followed by concluding remarks in Section 5.

2. Problem Description and Preliminaries

2.1. Problem Definition

A hybrid diesel/battery ship with a 1000 kW diesel generator and 300 kWh battery is considered. This ship sails on a special route involving both domestic and international waters. It is assumed that the ship can provide the possibility of charging the electric vehicles that it carries by itself. The maximum power transmission for charging electric vehicles is assumed to be constant and limited. The single line electric diagram of the ship is shown in Figure 1.

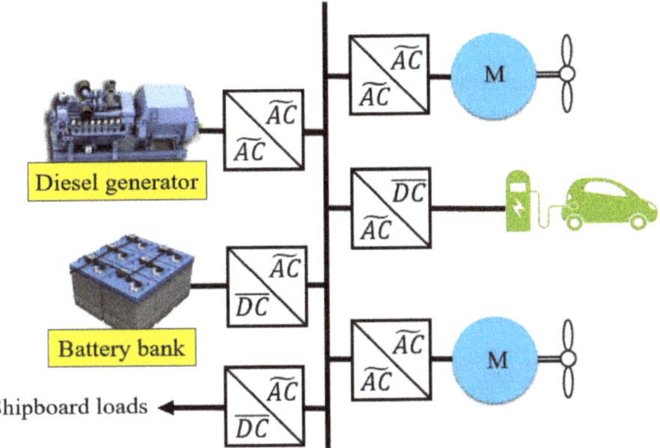

Figure 1. Single-line electric block diagram of the ship.

It should be noted that this type of ships usually sail a specific route two or three times per day. Energy management at each journey is performed considering discharging the batteries between predetermined values such that the total energy storage of the batteries are shared among all daily journeys. At the end of each day, after the last journey, the batteries are charged using cold-ironing in harbor. Since in this paper, the main focus is on only one journey, only the constraints related to discharging the batteries up to predetermined values at the end of understudy journey are considered in the modeling.

Basic parameters of the case study ship are presented in Table 1 and illustrated in Figure 2.

Table 1. Basic parameters of case study ship.

ρ_s (kg/m^3)	S (m^2)	L (m)	ϑ (m^2/s)	g (m/s^2)	∇ (m^3)	B (m)	T (m)	β
1.005	490.77	49.830	1.188×10^{-6}	9.8	485.5	11.297	1.920	0.7874

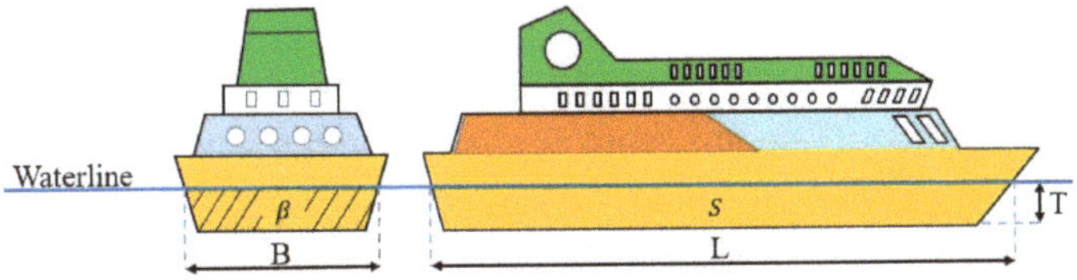

Figure 2. Illustrative representation of the understudy ship parameters.

It is assumed that the travelling starts from a country with specific emission restriction in its own domestic waters. No emission restrictions are considered for international waters. So, when the ship enters the international waters no emission restrictions is imposed to the ship operation. When the ship leaves the international waters and enters to the domestic waters of the destination country new emission control restrictions should be considered in the operation of the ship.

Different studies show that wind direction and sea state can affect the performance of ship electric motors [18]. It is assumed that the traveling route is short. So, the wind direction and condition of the sea in different traveling route locations can be predicted and known in the power scheduling problem.

The variations in ship's speed during travel lead to changes in ship's resistance. The consumed fuel is also characterized as a function of shipping speed and also ship's speed-dependence resistance. So, in order to determine the amount of fuel consumption, the relation between the ship's speed and resistance should be taken into account.

Considering all above mentioned assumptions, the aim of this paper is to determine the optimal speed adjustment of the ship during the journey and scheduling the batteries and electric vehicles charging considering the emission limitation, weather condition, and ship resistance such that the optimal operation cost is achieved.

It is assumed that energy resources scheduling is performed for different portions of the sailing route. Each portion is denoted by index x and the set of all portions is defined by X.

2.2. Ship's Resistance Formulation

Considering the ship's speed-dependent resistance and as mentioned in [19], the propulsion force is proportional to shipping speed as follows:

$$P_x{}^m = R(V_x) \times V_x \tag{1}$$

where $P_x{}^m$ denotes the electric motors' power in kW, $R(V_x)$ is the ship's speed-dependent resistance, and V is the shipping speed in (m/s). Ship's resistance is represented as:

$$R(V_x) = C_T(V_x)\frac{\rho_s}{2}V_x{}^2S \tag{2}$$

where ρ_s is mass density, S is the wetted surface of the ship, and the total resistance coefficient, denoted by C_T, is given by:

$$C_T(V_x) = C_F(V_x) + C_R(V_x) + C_A + C_{AA} + C_{AS}, \tag{3}$$

where C_F, C_R, C_A, C_{AA}, and C_{AS} represent the frictional resistance coefficient, the residual resistance coefficient, the incremental resistance coefficient, the air resistance coefficient, and the steering resistance coefficient, respectively. The frictional resistance coefficient, denoted by C_F, is formulated as:

$$C_F(V_x) = \frac{0.075}{\left(log_{10}^{R_n(V_x)} - 2\right)^2}, \tag{4}$$

R_n is the Reynolds Number and is defined as:

$$R_n(V_x) = \frac{V_x \times L}{\vartheta} \tag{5}$$

where L and ϑ are the length of the waterline and the coefficient of kinematic viscosity, respectively. C_R is expressed as a function of Froude number. The Froud number is defined as below:

$$F_n(V_x) = \frac{V_x}{\sqrt{g \times L}} \tag{6}$$

where g is the standard gravity. According the the available studies [19] the relation between C_R and Froude number is highly dependent on the parameters of the ship. In [19] relation between C_R and Froude number is depicted using some curves obtained by experimental results. In order to find the best curve for the ship under study, two parameters should be calculated. The first term is a length-displacement ratio that is defined as below:

$$LD = \frac{L}{\nabla^{\frac{1}{3}}} \tag{7}$$

The second term is the prismatic coefficient φ of the model given by:

$$\varphi = \frac{\nabla}{L \times B \times T \times \beta}, \tag{8}$$

where ∇, B, T, and β denote volumetric displacement, breadth, draught, and mid-ship section area coefficient, respectively. Using these two parameters the corresponding curve for C_R versus Froude number can be found in [19]. Curve fitting is applied to assign a mathematical formulation to the relation between C_R and fourth-degree. The following polynomial function represents this relation:

$$10^3 C_R(V_x) = (0.00942 \times {V_x}^4 - 0.1827 \times {V_x}^3 + 1.413 \times {V_x}^2 - 4.927 \times V_x + 6.75) \times 10^{-3} \quad (9)$$

Now, it should be noted that this curve is valid for the case that $\frac{B}{T} = 2.5$. Below formulation is used to find the value of C_R for other $\frac{B}{T}$ values:

$$10^3 C_R(V_x) = 10^3 C_{R,\ \frac{B}{T}=2.5}(V_x) + 0.16\left(\frac{B}{T} - 2.5\right) \quad (10)$$

The incremental resistance coefficient, C_A, depends on the length of the waterline and for understudy ship equals to 0.4×10^{-3}. It is also assumed that $C_{AA} = 7 \times 10^{-5}$, $C_{AS} = 4 \times 10^{-5}$. Mathematical representations of all resistant coefficients and the total resistant coefficient curve C_T are depicted in Figure 3. Ship's resistant is obtained by including the calculated value of C_T using (3), (4), (9), (10) and above-mentioned information.

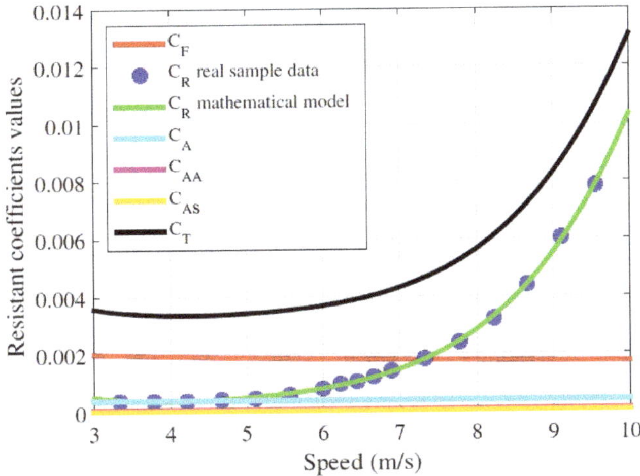

Figure 3. The resistance coefficients versus shipping speed.

2.3. Fuel Consumption

The following formulation is used to find the mass of fuel consumption for a specific output power of the diesel generator [20]:

$$M_x^F = SFOC \times {P_x}^{DG}, \quad (11)$$

where P^{DG} is the output power of the diesel generator in (kW) and M_x^F is the fuel consumption in (g/h). Specific fuel oil consumption (SFOC), in (g/kWh), is the measure of the amount of fuel consumed in (g) by the diesel generator to produce a unit of energy in (kWh). The SFOC at different loading of the diesel generator is derived from the performance curve, which can be obtained from the manufacturers. The SFOC curve is shown in Figure 4.

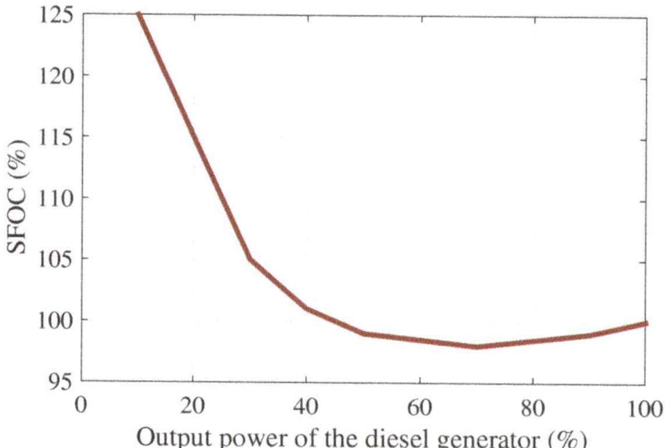

Figure 4. The SFOC curve for a diesel generator [20].

Considering that the SFOC in 100% output power of the power generator is 211 g/kWh [20] and capacity of diesel generator is 1000 kW, curve fitting using cftool in Matlab software is applied in Figure 4 to assign a mathematical function to the relation between the output power of the diesel generator and SFOC as below:

$$\begin{aligned} SFOC = & \, 2.04 \times 10^{-10} \times {P_x}^{DG^4} - 6.88 \times 10^{-7} \times {P_x}^{DG^3} \\ & + 8.689 \times 10^{-4} \times {P_x}^{DG^2} - 0.474 \times {P_x}^{DG^1} + 300.8 \end{aligned} \tag{12}$$

2.4. Impacts of Weather Condition on the Motors' Performance

One of the goals of this paper is considering the effects of weather condition on the optimal energy management of the ship. Reference [18] has already studied the impacts of wind direction and sea state on the fuel consumption efficiency of the ships. In [18], calculation are performed for ships with diesel engines whereas understudy ship in this paper is an all-electric ship that uses electric motors. However, since the fuel consumption and output power in diesel engines have a linear relation, same impact can be considered for wind direction and sea state on the efficiency of output power of the motors. So, obtained results in [18] are used in this paper to model the impact of wind speed and sea state on the efficiency of the motors. In the rest of this paper, the phrase "motors' efficiency" refers to the motors' electrical efficiency plus the ability of electric motors in reaching the ship's speed to a specific value. Analysis in [18] shows that tail wind, beam wind and head wind have different impacts on the fuel consumption of ships. Around the operation point of understudy system, tail wind and beam wind have similar impacts on the motors' efficiency and head wind reduces the motors' efficiency about 4%. Coefficient C_x^{WD} is introduced to consider the impact of wind direction on the motors' efficiency. We have:

$$C_x^{WD} = \begin{cases} 1 & \text{if } x \in X_{T,B} \\ 0.96 & \text{if } x \in X_H \end{cases}, \tag{13}$$

$X_{T,B}$ refers to the set of portions in which the wind blows to Tail or Beam of the ship and X_H is to the set of portions in which we have head wind.

The Beaufort scale is also used in [18] to categorize the different states of the sea and their impacts on the performance of motors. According to the Beaufort scale, 13 scales $\{0, 1, \ldots, 12\}$ are used to describe the sea state. Scale 0 represents the calm state and scale 12 defines a hurricane sea. In [18], the focus is on three states 4, 5, and 6. In state 4, wind speed is between 20 to 28 km/h and wave height is between 1 to 2 m. In state 5, wind speed is between 29 to 38 km/h and wave height is between 2 to 3 m, and in state 6, wind speed is between 39 to 49 km/h and wave height is between 3 to 4 m. Based on the analysis in [18], considering the state 4 as the base case, in states 5 and 6 the motors' efficiency decreases about 4% and 8%, respectively. Coefficient C_x^{SS} is introduced to consider the impact of see state on the motors' efficiency as below:

$$C_x^{SS} = \begin{cases} 1 & \text{if } x \in X_{SS4} \\ 0.96 & \text{if } x \in X_{SS5}, \\ 0.92 & \text{if } x \in X_{SS6} \end{cases} \tag{14}$$

Sets X_{SS4}, X_{SS5}, and X_{SS6} refer to sets in which the sea state is 4, 5, and 6, respectively. Now, using coefficients C_x^{WD} and C_x^{SS} the motors efficiency is formulated as below:

$$ef_m = ef_{ele} \times C_x^{WD} \times C_x^{SS} \tag{15}$$

ef_{ele} represents the motors' electrical efficiency. It should be noted that since the traveling time period is short, it is assumed that wind direction and sea state are known at the beginning of the sailing.

3. Optimization Problem

In this section, the optimal scheduling problem of the hybrid diesel/battery ship is formulated. The goal is to minimize the total shipping operation cost considering diesel generator's operation constraints, the battery's operation constraints, time limitation, the fuel consumption limit, and emission constraint. The whole sailing route is divided into several portions. It is assumed that the adjusted variables are constant during each portion of the trip. The problem is formulated as below:

$$\min \sum_{x \in X} (C^F M_x^F + C^{BD} P_x^{Dch} - C^{EV} P_x^{EV-ch}) t_x + K \times \sum_{x \in X-\{1\}} (V_x - V_{x-1})^2 \tag{16}$$

$$\text{s.t.} \quad t_x = \frac{D_x}{3.6 \times 10^3 V_x}, \tag{17}$$

$$M_x^F = \frac{P_x^{DG}}{3.6 \times 10^6} \left(2.04 \times 10^{-10} \times P_x^{DG^4} - 6.88 \times 10^{-7} \times P_x^{DG^3} + 8.689 \times 10^{-4} \times P_x^{DG^2} - 0.474 \times P_x^{DG^1} + 300.8\right) \tag{18}$$

$$P_x^m = R_T V_x = 246 \times \left(\frac{0.075}{(log_{10}^{41.94 \times 10^{-6} V_x} - 2)^2} + (0.00942 \times {V_x}^4 - 0.1827 \times {V_x}^3 + 1.413 \times {V_x}^2 - 4.927 \times {V_x}^1 + 6.75 + 0.54) \times 10^{-3} + 0.4 \times 10^{-3} + 0.07 \times 10^{-3} + 0.04 \times 10^{-3}\right) \times {V_x}^3 \tag{19}$$

$$0 \le P_x^{DG} \le P^{DG,max} \tag{20}$$

$$P_x^{DG} - P_{x-1}^{DG} \le R^{max} \tag{21}$$

$$\underline{SOC^B} \le \frac{E_x^B}{B^{max}} \le \overline{SOC^B} \qquad \forall\, x \in X \tag{22}$$

$$P_x^{Dch} \le R^{Dch} \qquad \forall\, x \in X \tag{23}$$

$$Y_x^{dc} = 1 \; if \; P_x^{Dch} \ge 0 \tag{24}$$

$$E_x^B = E_{x-1}^B - \left[ef^{ch}(1 - Y_x^{dc}) + \frac{1}{ef^{dc}} Y_x^{dc}\right] \times P_x^{Dch} t_x \tag{25}$$

$$E_0^B = E_0, \ E_{N_x}^B = E_f \tag{26}$$

$$\sum_{x \in X} t_x = \bar{T} \tag{27}$$

$$\sum_{x \in X} M_x^F t_x \leq M^{max} \tag{28}$$

$$\frac{F \times M_x^F}{M_{cargo} \times D_x} \leq EEOI_x^{max,A} \quad \forall \, x \in X_A \tag{29}$$

$$\frac{F \times M_x^F}{GT \times D_x} \leq EEOI_x^{max,B} \quad \forall \, x \in X_B \tag{30}$$

$$P_x^{EV-ch} < P_x^{EV-max} \tag{31}$$

$$\sum_{x \in X} P_x^{EV-ch} t_x = E^{EV-max} \tag{32}$$

$$P_x^{DG} + P_x^{Dch} = P_x^{load} + \frac{P_x^m}{ef_m} + P_x^{EV-ch} \tag{33}$$

The first term of objective function (16) is the cost of fuel consumption. C^F and M_x present the fuel price in ($\frac{\$}{g}$) and the consumed mass of fuel in ($\frac{g}{h}$), respectively. The second term in (16) denotes a cost that refers to batteries degradation due to discharging. C_{BD} is its price and P_{BD} is the discharged power of the batteries in portion x. As mentioned before, it is assumed that the ship provides the possibility of charging electric vehicles. The third term is the total income for charging electrical vehicles on board C^{EV} and P^{EV-ch} are the price and power of charging electrical vehicles at the portion x, respectively. t_x is the traveling time at portion x in (h). The last term in the objective function is a penalty that prevents fast variations and fluctuations in the speed of the ship during the sailing. Parameter K denotes the importance of this penalty function in the optimization.

Constraint (17) represents the relation between speed and time period of each portion. Constraint (18) defines the relation between fuel consumption and output power of diesel power generator using Equations (11) and (12). Equation (19) denotes the relation between motors' power, speed, resistant, an weather condition.

Constraint (20) corresponds to the maximum output power of diesel generator where $P^{DG,max}$ is the rated power of the diesel generator. Changing the operation point of the diesel generator happens at the beginning of the portion x and is limited. Constraint (21) limits the variation of the diesel generator's output power when the ship arrives at the portion x.

Constraints (22)–(26) are related to the batteries operation limitations. $\underline{SOC}^B$ and $\overline{SOC}^B$ are upper and lower bounds of the batteries state of charge (SOC), respectively. The charging and discharging power of the batteries is limited to R^{Dch} in (23). Constraint (24) is used to determine the charging or discharging state of the batteries. Adjusting the zero or one value to Y_x^{dc} helps us to assign charging or discharging efficiencies to the batteries operation in constraint (25). The equality constraint (25) is used to update the energy storage of the batteries at the end of each portion of the trip. Constraints in (26) assign the initial and final energy storage of the batteries.

Traveling time period is limited by (27). Total available mass of fuel is limited to M^{max} in (28).

In this paper, it is assumed that the emission control policies are various in different portions of the sailing route. Sets X_A and X_B refer to the portions close to the start and ending points of the sailing routes, respectively. The maximum value for the emission index in sets X_A and X_B are different. Constraints (29) and (30) are defined to illustrate these different emission control policies. In (29) and (30), F is the coefficient for converting mass of fuel to CO_2, GT represents the gross tonnes of the ship, and $EEOI_x^{max,A}$ and $EEOI_x^{max,B}$ denote the emission limits in the areas of sets X_A and X_B. No emission restriction is considered for other portions of the sailing route.

Constraint (31) limits the power transfer for charging the electric vehicles to P_x^{EV-max} and constraint (32) denotes the maximum free energy storage capacity in the candidate electric vehicles in ship for charging, i.e., E^{EV-max}.

Constraint (33) represent the power generation and consumption balance of the ship. Parameter P_x^{load} denotes the shipboard loads except the motors power in portion x.

Proposed optimization problem (16)–(33) is non-linear model. The Genetic Algorithm (GA) toolbox in MATLAB software is applied to solve this optimization problem.

4. Simulation Results

In this section, the proposed energy management strategy is applied to the real case study ship introduced in Section 2 and Table 1. Other required information for the simulation are presented in Table 2. Shipboard loads are presented in Figure 5. The understudy ship is assumed to operate on the Sweden-Denmark ferry route Goteborg-Frederikshavn. The length of the route is about 110 km and the crossing time is 5 h. The sailing route is divided into 15 portions and the length of each portion is 7.33 km.

Table 2. Parameters of the energy system.

Parameter	Value
Rated power of the diesel generator ($P^{DG,max}$)	1000 kW
Power ramp rate of the diesel generator (R^{max})	250 kW
Total power of electric motors (P_e^{max})	1000 kW
The efficiency of electric motors (ef_{ele})	0.85
The batteries capacity (B^{max})	300 kWh
Charge and discharge rate of the batteries	50 kW
Maximum SOC of the batteries ($\overline{SOC}^B$)	0.8
Minimum SOC of the batteries ($\underline{SOC}^B$)	0.2
Charging efficiency of batteries (ef^{ch})	0.85
Discharging efficiency of batteries (ef^{ch})	0.95
Fuel consumption to CO_2 conversion coefficient (F)	3.2
Gross tonnes of the ship (GT)	650 t
$EEOI_x^{max,A}$ for Denmark territorial waters	23 $\frac{tCO_2}{t \times Nm}$
$EEOI_x^{max,B}$ for Sweden territorial waters	26 $\frac{tCO_2}{t \times Nm}$
Power transmission limit for charging the EVs (P_x^{EV-max})	20 kW
Total required energy for charging the EVs (E_x^{EV-max})	70 kWh
Fuel tank capacity (M^{max})	1500 kg
Weighting coefficient of speed variations penalty function (K)	50

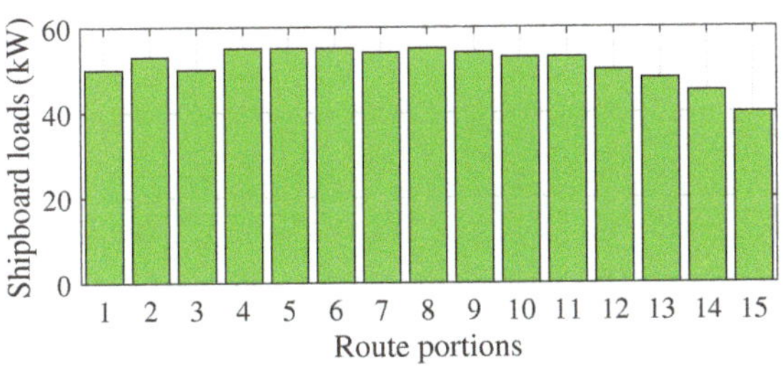

Figure 5. Shipboard load at different route portions.

According to the 1982 United Nations Convention on the Law of the Sea, the territorial sea of each country is a belt of coastal waters extending at most 22 km from the baseline of a coastal state. Since it is assumed that the emission controls are for the territorial waters of each country and the length of each portion is 7.33 km, the first three portions of the route are assigned to the emission control area of Denmark and last three portions of the route are assigned to the emission control area of Sweden.

Wind direction and sea state at each portion are presented in Table 3. In Table 3, symbols TB and H represent the tail or beam wind and head wind respectively, and symbols S4, S5, and S6 represent the sea states 4, 5, and 6 respectively. According to (13) and (14) and information in Table 3, values of motors' efficiency in (15) at each portion of the route is calculated as shown in Figure 6. As shown in Figure 6, in first seven portions the weather condition affects the motors' efficiency less than the last eight portions of the sailing route. In next subsections, first, the simulation results of the proposed model (16)–(33) that is refereed as the base case is presented and compared with the case that no emission policy restrictions are considered. Then, impacts of weather condition on the simulation results are investigated and operation costs at different operation cases are compared.

Table 3. Wind direction and sea state in different portion of the trip.

Portion	1	2	3	4	5	6	7	8	9	10	11	12	13	14	15
Wind direction	H	H	H	TB	TB	TB	TB	TB	TB	H	H	H	H	H	H
Sea state	S4	S4	S4	S4	S4	S5	S5	S6	S6	S6	S6	S6	S6	S6	S5

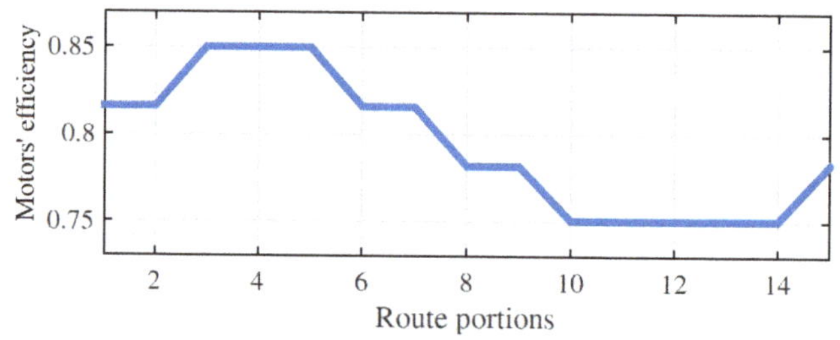

Figure 6. Impacts of weather condition on the motors' efficiency.

4.1. Simulation Results of the Proposed Energy Management Method

Optimal scheduling results of different variables are presented in Figures 7–11. In Figure 7, the speed is presented in Knot unit that is more common unit for speed measurement at seas. We have $V(Knot) = 0.5144 \times V(\mathrm{m/s}) = 1.852 \times V(\mathrm{km/h})$. As shown in Figure 7, emission limitations impose reducing the ship' speed in first and last three portions of the route. Since the emission limitation in first three portions are more than emission limitations in the last three portions, the speed in first three portion is scheduled less than the speed in the last three portions. In international waters, i.e., portions 4–12, the ship's speed changes such that the fast variations in the speed is prevented, the power ramp rate constraint of the diesel generator is satisfied and the ship arrives to the distention on time. This has let to an almost symmetrical scheduling for the speed in these portions.

Scheduled output powers of the diesel generator are presented in Figure 8. According to Figure 8 diesel generator's output power has a trend similar to the trend of speed due to the direct dependency of speed to power. In international waters, produced energy by the diesel generator in the First 50% of the route is about 6.5% less than its produced energy in the next 50% of the route. This happens because in the First 50% of the route the efficiency of the motors are high and hence, less energy is needed to reach to the scheduled speed. Power ramp rate constraint of the diesel generator is activated in portions 12 and 13.

Figure 9 compares the batteries' operation in the base case and the case that emission restrictions are not considered. When emission restrictions are ignored the batteries are discharged step by step during the sailing, but when the emission restriction are considered the battery is discharged mostly in first and last three portions. In fact batteries' energy is discharged in these portions for supplying the motors instead of producing energy by diesel generator to meet the emission policy restrictions and also arrive to the destination on time.

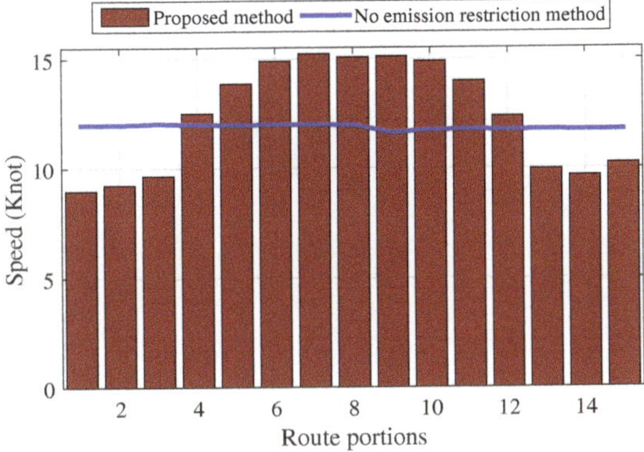

Figure 7. Ship speed scheduling results.

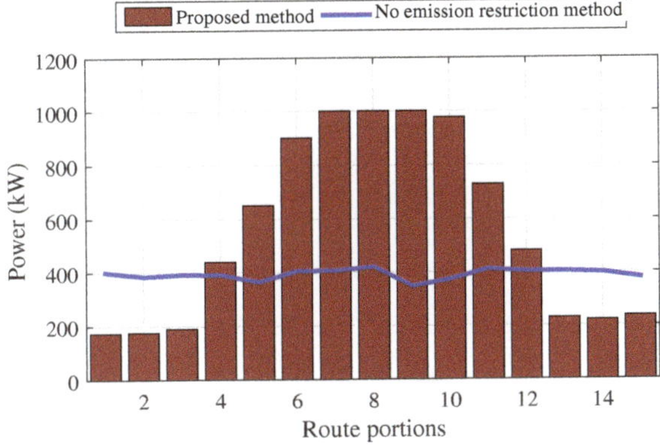

Figure 8. Scheduling results of the diesel generator's output power.

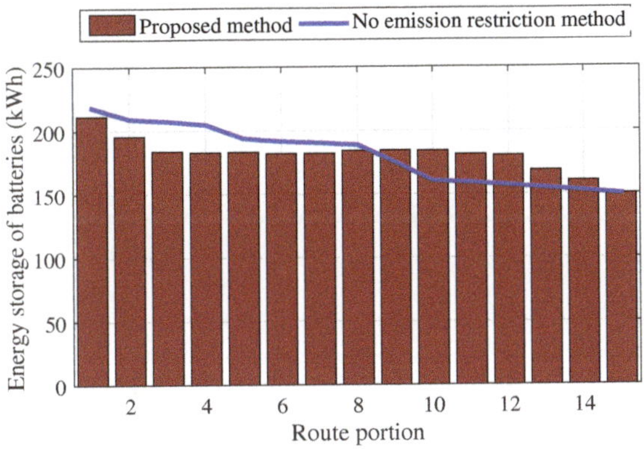

Figure 9. Energy storage of batteries at different route portions.

According to Figure 10, EVs are mostly charged when the ship is in international waters and sail in a stable speed. The average charged power of the EVs reduces when the ship is in the areas with emission restriction and when the speed changes rapidly.

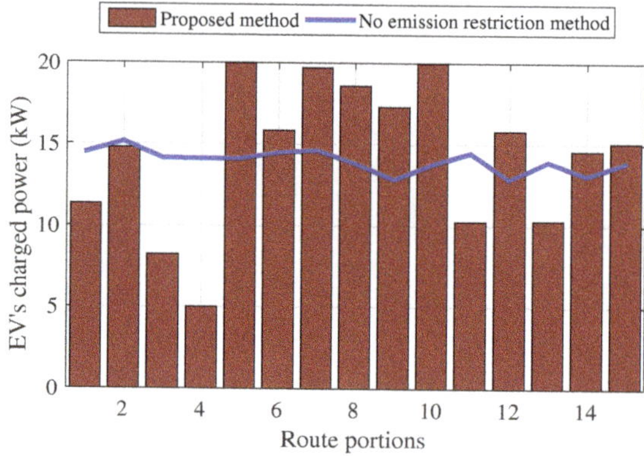

Figure 10. Scheduling results of EVs' charged power at different route portions.

Values of the emission index used in (29) and (30) are presented in Figure 11. As shown in Figure 11, emission is controlled in the first and last three portions of the route compared to the case that emission restrictions are not considered in the model. Figure 11 also shows that for the understudy system, considering the emission restrictions in some areas has led to increasing the total value of emission index over all route portions about 34%. So, it can be concluded that local emission limitations might help for reducing the emission in specific areas but increases the emission in overall which is not proper environmentally.

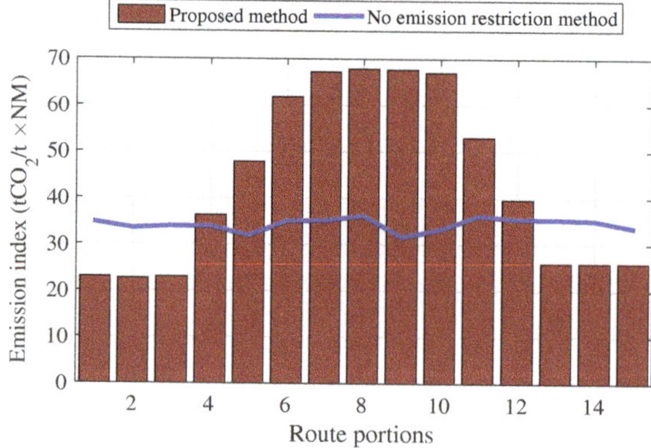

Figure 11. Emission index values at different route portions.

4.2. Impacts of Weather Condition on the Results

In order to study the impacts of weather conditions on the results, a simulation is performed after removing equations related to weather conditions, i.e., Equations (13)–(15) from the optimization problem. Simulation results are presented in Figures 12–14. As shown in Figures 12 and 13, ignoring the weather conditions in the formulation results less estimated generated power by diesel generator (dashed blue line) and more estimated speed (blue line) for the ship in some portions when the motors' efficiency decreases in the base case. This is obviously the result of ignoring the wind direction and sea state in the model. However, although these impacts are ignored in the mathematical formulation, they will affect the ship operation in practice. In fact, in order to reach to the scheduled analytical speed (blue line) in Figure 13, the actual required output power for diesel generator will be similar to the Figure 12 (pink line). But this power is above the maximum output power of the diesel generator and the output power will be set on the maximum value in these portions (green line). This leads to reduction in the speed of the ship in these portions of the route and could lead to delay on arriving the ship to its destination.

The charging and discharging scheduling of the batteries is also affected by ignoring the impacts of weather conditions in the optimization problem. Figure 14 compares the charging and discharging scheduling of the batteries in the case of ignoring the impacts of weather condition with the base case. According to Figure 14, while in the base case the most discharged power of the batteries are used in first and last three portions, the batteries are discharged almost uniformly in the case that the impacts of weather condition are ignored. This difference is explained as follows. The output power of diesel generator in both cases is limited in the first and last three portions because of the emission restrictions in the first and last three portions of the route. Now, due to the impacts of weather condition on the motors' efficiency more required power is calculated for the motors in the base case compared to the case that the impacts of weather condition are ignored. Since, the output power of the diesel generator is limited in both cases due to emission restrictions, this extra required in energy in the base case in the first and last three portions is obtained from the batteries. This leads to more discharged power of batteries in the first and last three portions at the base case compared to the case that the impacts of weather condition are ignored.

It should be noted that in this paper only the impacts of three consecutive sea state conditions out of thirteen different conditions are considered in the model. Considering more states can highlight the impacts of weather conditions on the energy management results significantly.

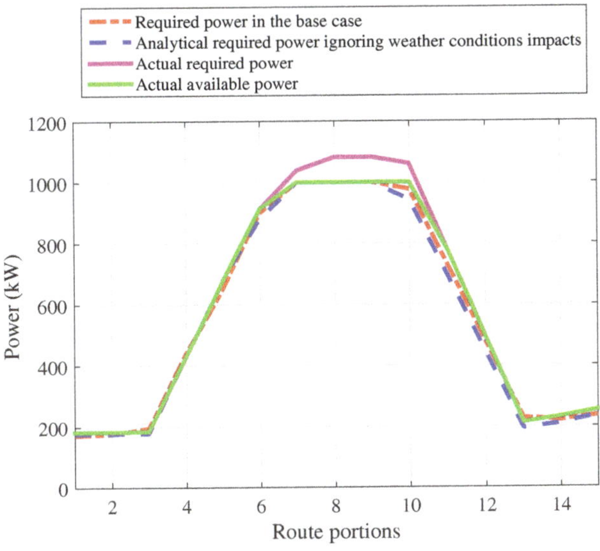

Figure 12. Comparing the output power of the diesel generator with and without considering the impacts of weather conditions.

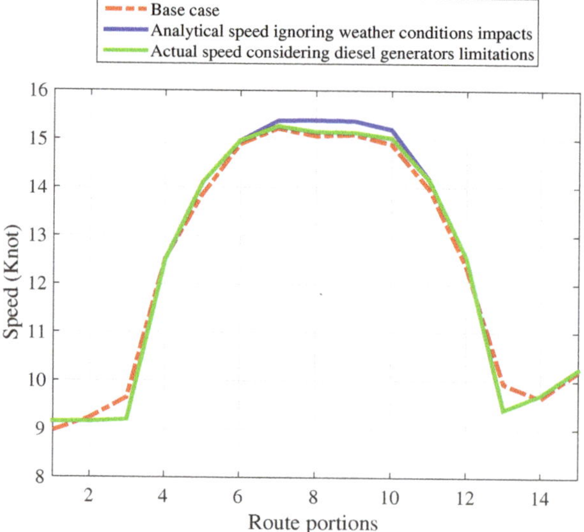

Figure 13. Comparing the speed of the ship at different portions of the route with and without considering the impacts of weather conditions.

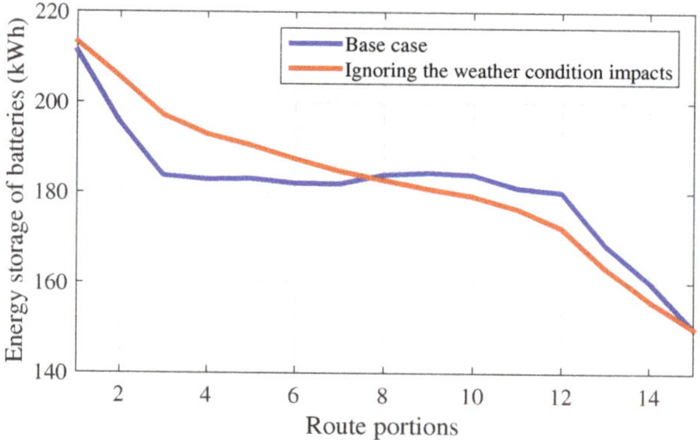

Figure 14. Comparing the energy storage of the batteries with and without considering weather impacts at different route portions.

4.3. Comparing the Operation Cost of the Ship in Different Cases

In this subsection, the ship's operation cost is investigated in different cases. Understudy cases are as follows:

Case 1: fixed speed, discharged power of batteries, and charged power in EVs in all portions;
Case 2: optimal resources scheduling without emission policy restrictions;
Case 3: optimal resources scheduling without considering weather condition; and
Case 4: optimal resources scheduling considering weather conditions and emission restrictions.

Simulation results are presented in Table 4. As shown in Table 4, case 2 is the least fuel consumption and operation cost case in which there is no emission limitation and also the impacts of weather conditions are considered in the model. Case 1 that is the most common method for supplying the resources in the ship is about 0.8% more expensive because of ignoring the impacts of weather condition. When the emission restrictions are considered in the model (case 4), the operation cost increases about 28%. If the weather conditions are ignored in the formulation (case 3), analytical operation cost obtains USD 782 but in practice the operation cost will be USD 833.69 due to impacts of weather conditions on the realized operation of the ship. This means that considering the weather condition in the formulation can reduce the total operation cost up to 3.5%.

Table 4. Comparing the fuel consumption, emission, and operation cost in different cases.

	Fuel Consumption (Liter)	Total Emission Index Violations (tCO_2/t*NM)	Total Operation Cost ($)
Case 1	502.8	59.28	634.54
Case 2	498.4	59.62	629.55
Case 3	655.4	3.5	833.69
Case 4	633.12	0	804.81

5. Conclusions

In this paper, cost effective energy management problem of a hybrid diesel/battery ship that sails in areas with different emission control policies is studied. The understudy energy system utilizes diesel generator and batteries as the electric energy resources. Ships' electric motors, shipboard loads, and electric vehicles that need to be charged while carrying by the ship are the main electric loads. The proposed method formulates the relation between the speed and the resistant of the understudy ship. This helps to find an accurate model for determining the relation between the required propulsion power and the speed of the ship. Moreover, impacts of wind direction and sea state on the efficiency of the ship's electric motors are formulated and involved in the model. A non-linear optimization model is proposed for the understudy energy system and the genetic algorithm is used to solve the problem. The proposed method is applied to a real case ship. Simulation results confirms the effectiveness of the proposed method in minimizing the operation cost, satisfying different operation constraints, and providing the required energy for charging the electric vehicles while considering the emission restrictions in different areas. Simulations results shows that considering the impacts of weather condition on the results can reduce the operation cost of the ship up the 3.5% and ignoring the impacts of weather condition not only affect the scheduling of the batteries significantly but also leads to delay on arriving to the destination on time. Simulation results also shows that while limiting the CO_2 emission in some areas satisfies the emission restrictions in that areas, it can lead to increasing the total produced CO_2 in overall which is not proper environmentally.

Author Contributions: Investigation, M.B.; methodology, M.B., F.G. and M.R.; software, J.B.; supervision, M.-H.K. All authors have read and agreed to the published version of the manuscript.

Funding: This work was supported by the Energy Technology Development and Demonstration Program (EUDP) under Grant (64018-0721), HFC: Hydrogen Fuel Cell and Battery Hybrid Technology for Marine Applications.

Conflicts of Interest: The authors declare no conflict of interest.

References

1. Rafiei, M.; Boudjadar, J.; Khooban, M.H. Energy Management of a Zero-Emission Ferry Boat with a Fuel Cell-Based Hybrid Energy System: Feasibility Assessment. In *IEEE Transactions on Industrial Electronics*; IEEE: Piscataway, NJ, USA, 2020.
2. Letafat, A.; Rafiei, M.; Ardeshiri, M.; Sheikh, M.; Banaei, M.; Boudjadar, J.; Khooban, M.H. An Efficient and Cost-Effective Power Scheduling in Zero-Emission Ferry Ships. *Complexity* **2020**, *2020*, 6487873. [CrossRef]
3. Hasanvand, S.; Rafiei, M.; Gheisarnejad, M.; Khooban, M.H. Reliable Power Scheduling of an Emission-Free Ship: Multi-Objective Deep Reinforcement Learning. *IEEE Trans. Transp. Electrif.* **2020**, *6*, 832–843. [CrossRef]

4. Khooban, M.H.; Gheisarnejad, M.; Farsizadeh, H.; Masoudian, A.; Boudjadar, J. A new intelligent hybrid control approach for DC–DC converters in zero-emission ferry ships. *IEEE Trans. Power Electron.* **2019**, *35*, 5832–5841. [CrossRef]
5. Vafamand, N.; Boudjadar, J.; Khooban, M.H. Model predictive energy management in hybrid ferry grids. *Energy Rep.* **2020**, *6*, 550–557. [CrossRef]
6. Geertsma, R.; Negenborn, R.; Visser, K.; Hopman, J. Design and control of hybrid power and propulsion systems for smart ships: A review of developments. *Appl. Energy* **2017**, *194*, 30–54. [CrossRef]
7. Barelli, L.; Bidini, G.; Gallorini, F.; Iantorno, F.; Pane, N.; Ottaviano, P.A.; Trombetti, L. Dynamic Modeling of a Hybrid Propulsion System for Tourist Boat. *Energies* **2018**, *11*, 2592. [CrossRef]
8. Tang, R.; Li, X.; Lai, J. A novel optimal energy-management strategy for a maritime hybrid energy system based on large-scale global optimization. *Appl. Energy* **2018**, *228*, 254–264. [CrossRef]
9. Kanellos, F. Optimal power management with GHG emissions limitation in all-electric ship power systems comprising energy storage systems. *IEEE Trans. Power Syst.* **2013**, *29*, 330–339. [CrossRef]
10. Hou, J.; Song, Z.; Park, H.; Hofmann, H.; Sun, J. Implementation and evaluation of real-time model predictive control for load fluctuations mitigation in all-electric ship propulsion systems. *Appl. Energy* **2018**, *230*, 62–77. [CrossRef]
11. Li, Z.; Xu, Y.; Fang, S.; Wang, Y.; Zheng, X. Multi-objective Coordinated Energy Dispatch and Voyage Scheduling for a Multi-energy Ship Microgrid. *IEEE Trans. Ind. Appl.* **2019**, *56*, 989–999. [CrossRef]
12. Balsamo, F.; Capasso, C.; Miccione, G.; Veneri, O. Hybrid storage system control strategy for all-electric powered ships. *Energy Procedia* **2017**, *126*, 1083–1090. [CrossRef]
13. Paran, S.; Vu, T.; El Mezyani, T.; Edrington, C. MPC-based power management in the shipboard power system. In Proceedings of the 2015 IEEE Electric Ship Technologies Symposium (ESTS), Alexandria, VA, USA, 21–24 June 2015; pp. 14–18.
14. Han, J.; Charpentier, J.F.; Tang, T. An energy management system of a fuel cell/battery hybrid boat. *Energies* **2014**, *7*, 2799–2820. [CrossRef]
15. Letafat, A.; Rafiei, M.; Sheikh, M.; Afshari-Igder, M.; Banaei, M.; Boudjadar, J.; Khooban, M.H. Simultaneous energy management and optimal components sizing of a zero-emission ferry boat. *J. Energy Storage* **2020**, *28*, 101215. [CrossRef]
16. Bassam, A.M.; Phillips, A.B.; Turnock, S.R.; Wilson, P.A. Development of a multi-scheme energy management strategy for a hybrid fuel cell driven passenger ship. *Int. J. Hydrog. Energy* **2017**, *42*, 623–635. [CrossRef]
17. Banaei, M.; Rafiei, M.; Boudjadar, J.; Khooban, M.H. A comparative analysis of optimal operation scenarios in hybrid emission-free ferry ships. *IEEE Trans. Transp. Electrif.* **2020**, *6*, 318–333. [CrossRef]
18. Bialystocki, N.; Konovessis, D. On the estimation of ship's fuel consumption and speed curve: A statistical approach. *J. Ocean. Eng. Sci.* **2016**, *1*, 157–166. [CrossRef]
19. Kristensen, H.O.; Lützen, M. Prediction of resistance and propulsion power of ships. *Clean Shipp. Curr.* **2012**, *1*, 1–52.
20. Yuan, L.C.W.; Tjahjowidodo, T.; Lee, G.S.G.; Chan, R.; Ådnanes, A.K. Equivalent consumption minimization strategy for hybrid all-electric tugboats to optimize fuel savings. In Proceedings of the 2016 American Control Conference (ACC), Boston, MA, USA, 6–8 July 2016; pp. 6803–6808.

MDPI
St. Alban-Anlage 66
4052 Basel
Switzerland
Tel. +41 61 683 77 34
Fax +41 61 302 89 18
www.mdpi.com

Energies Editorial Office
E-mail: energies@mdpi.com
www.mdpi.com/journal/energies

www.ingramcontent.com/pod-product-compliance
Lightning Source LLC
LaVergne TN
LVHW070651170726
843515LV00004B/385

* 9 7 8 3 0 3 6 5 1 7 1 1 7 *